W9-BCF-753

LIVING ICE

A view northeastward up Barnard Glacier in Alaska's St. Elias Mountains. The dark stripes (medial moraines) show that this compound valley glacier consists of at least a dozen principal ice streams. The large central medial moraine forms at the junction of the two major branches in the middistance. Note how tributaries from the right make room by boldly pushing into the trunk glacier. Such ice streams would fare well on crowded subway cars – they play rough! (August 20, 1949.)

LIVING ICE

Understanding Glaciers and Glaciation

ROBERT P. SHARP
California Institute of Technology

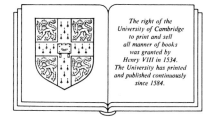

The right of the
University of Cambridge
to print and sell
all manner of books
was granted by
Henry VIII in 1534.
The University has printed
and published continuously
since 1584.

Cambridge University Press

Cambridge

New York   Port Chester   Melbourne   Sydney

Published by the Press Syndicate of the University of Cambridge
The Pitt Building, Trumpington Street, Cambridge CB2 1RP
40 West 20th Street, New York, NY 10011, USA
10 Stamford Road, Oakleigh, Melbourne 3166, Australia

© Cambridge University Press 1988

First published in 1988
First paperback edition 1991

Printed in the United States of America

Library of Congress Cataloging-in-Publication Data

Sharp, Robert P. (Robert Phillip)
Living ice: understanding glaciers and glaciation / Robert P. Sharp.
p. cm.
Bibliography: p.
Includes index.
ISBN 0-521-33009-2 (hc)
1. Glaciers 2. Glacial landforms I. Title
GB2403.2.S5 1988   88-3800
551.3′1–dc19   CIP

British Library Cataloging in Publication Data

Sharp, Robert P.
Living ice.
1. Glaciers. Physical properties
I. Title
551.3′12

ISBN 0-521-33009-2 hardback
ISBN 0-521-40740-0 paperback

To Walter A. Wood,
and the memory of Foresta H. Wood and Valerie F. Wood,
who so generously and joyfully introduced me to the world of
living glaciers.

Contents

Contents

Preface

Among scientists geologists get as much enjoyment out of their profession as any, especially while traveling. They see the world through a special pair of glasses that makes it an endlessly interesting and entrancing place. One objective of this small volume is to share with its readers some of those pleasures and enjoyments within two subdivisions of the earth sciences, **glaciology*** (the study of glaciers) and **glacial geology** (the tasks performed by glaciers).

Everyone knows that a baseball game is more interesting if the spectators are personally involved through knowing the names of the players and something about their abilities, accomplishments, and even their idiosyncrasies – Willie Mays' basket catch or Pete Rose's head-first slide, for example. It is the same with science. We need to know and understand some basic principles in order to enjoy the game. This volume endeavors to impart such knowledge in an understandable and uncomplicated way.

People are attracted by activities associated with construction of new buildings; hence sidewalk superintending is a time-honored avocation. With knowledge and an adventuresome spirit, we can do some sidewalk superintending of glaciers at work; if not in fact, then in our imagination. **Geology** is often concerned with events that happened in the past, and that is true of glacial geology, which deals principally with the erosional and depositional work of glaciers long gone. Glaciology, by contrast, deals with glaciers actually at work. That can be exciting because we can watch the game being played, or failing that, we can see it through the eyes of others who have. Glaciers work rapidly enough for results to be apparent within a few days or months.

Once we understand how glaciers work, our ability to comprehend what they have done in the past and are capable of doing in the future is enhanced. The landscape created by past glaciations becomes an open

---

* Terms in **boldface type** are defined in the Glossary beginning on page 197.

book for our reading. An objective of this dissertation is to provide information that will lead to an understanding of glacial processes.

This is not an all-about book. Its purpose is to touch upon some basic aspects of glaciers, their behavior, and the principal ways they shape the landscape. Information is presented in an informal, conversational manner, with the aim of exciting laypersons' interest in natural phenomena that professionals find fascinating.

I am deeply beholden to Enid H. Bell and Paul Hawley for services of typing, editing, and so expertly attending my manuscript needs, and to Dorothy Coy McDougall for backup typing and administrative assistance. Peter-John Leone has proved a valued colleague for his guidance, suggestions, and critical editorial comments, and Alan Gold has done a thoroughly professional job of copyediting. The many people providing photographs, individually identified in figure captions, are graciously thanked, with a bow to David R. Hirst for special help. All unacknowledged photographs are by Sharp. My greatest debt is to all those people, too numerous to specify, who have studied glaciers and their products and have shared their findings by publication.

Silver Gate, Montana

LIVING ICE

I

Glaciers and how they are made

Glaciers are active creatures, delicately attuned to their environment. They can be brutally overpowering yet surprisingly sensitive to subtle influences when exercising their strength. They expand and shrink, advance and recede, enjoy robust health and suffer unsightly deterioration. Like humans, they win victories and suffer defeats. They work diligently and vigorously at their tasks and display more common sense, economically, than most modern governments in maintaining balanced budgets. Glaciers prefer to inhabit cold, wet places not heavily populated or much visited by people. Some, however, exist in accessible mountainous terrains of more temperate regions where they contribute significantly to the dazzling scenery. Glaciers can be dramatically beautiful and are almost always interesting because of their activity; seldom are they inert. They can, however, be deadly for the unwary. Crevasses and avalanches are abundant, and both have taken a toll of human life.

Glaciers are important for what they are currently doing to the earth's surface, for their influence on local and worldwide climate, for their control of sea level, and particularly for what they have done in the recent past when as much as 32 percent of the land was covered by glacier ice.

AN IMAGINARY COMMISSION

Let us assume we have been directed by some agency, foundation, or supreme being to create a glacier and given authority to requisition substances and environments necessary to carry out the task. Our first request would probably be for **snow**,* as much snow as possible. That is a natural way to start; however, many locations on Earth receive huge amounts of snow each year but have no glaciers, and other places with

* Terms in **boldface type** are defined in the Glossary beginning on page 197.

light snowfalls have massive glaciers. Clearly, there is another all-important factor in the game of glacier creation, and it is *conservation*. Some of that snow must be saved, year after year after year, if we hope to make a glacier. Indeed, the two largest ice masses on Earth, the Antarctic and Greenland ice sheets, exist in regions of minimal precipitation. These great ice masses prosper because they hoard every flake of the scanty snow that falls; they are thrifty.

WHAT IS A GLACIER?

It seems reasonable at this point to determine just what we have been directed to create; we need to know what is and what is not a glacier. In a few words, a **glacier** is a natural body of ice, originating on land, and undergoing movement that transports ice from an area of accumulation to an area of disposal. To understand thoroughly the task ahead of us, we need to expand upon this definition by discussing more fully some of its implications.

First, a glacier must be of natural origin. Because we were commissioned to *make* a glacier, our project might appear to be doomed by the artificiality of the product. We are going to employ nature to do the job, however, using natural materials within natural environments, so the result should be acceptable. Second, to qualify, a glacier must form on land, although subsequently it may go to sea as an **iceberg**, like the one that sank the *Titanic*, or as an **ice shelf**, such as the huge Ross Ice Shelf of Antarctica.

Not all masses of ice of natural origin and associated with land are glaciers. Glaciers are dynamic entities engaged in accumulating, transporting, and disposing of ice. Transportation is a critical function. A mass of ice has to be able to move, in part by mechanisms of internal adjustment, to qualify as a glacier. A block of ice broken from an ice field and sliding down a mountainside is a moving mass of ice, but it is not a glacier. Bodies of **ground ice** formed by freezing of subsurface waters within **perennially frozen ground** are not glaciers, nor is the sea ice (Fig. 1.1) of the Arctic and Antarctic oceans. **Sea ice** is formed as a relatively thin layer on the ocean surface primarily by the freezing of seawater. Sea ice may contain drifting **ice islands** as much as 100 meters thick that have broken free from land-born ice shelves (Fig. 1.2), but sea ice itself does not qualify as a glacier. Although a glacier on the Antarctic Continent buries the South Pole under 2,800 meters of ice,

*Figure 1.1. Edge of the arctic sea ice off the coast of Greenland (background). Individual plates within or detached from the continuous sheet are called **floes**, not icebergs. The sea ice is frozen seawater, not glacier ice. Icebergs derived from glaciers and caught within the pack project above its surface in the lower right and farther afield (arrows). Sea ice is mostly 3 to 4 meters thick; icebergs are much thicker and hence float higher. (U.S. Army Air Force photo [10R-194][2-54].)*

there are no glaciers at the North Pole, only sea ice floating on the Arctic Ocean.

Finally, we need to recognize that glacier ice is a type of rock, one that covers a larger fraction of Earth's land surface than any other single species of rock. A **rock** is a solid natural body consisting of an aggregate of one or more minerals in the form of particles or crystals. A **mineral** is a natural inorganic solid of fixed chemical composition and definite

Figure 1.2. The white line curving gracefully across the lower part of this picture is an ice cliff marking the edge of a floating ice shelf off the north coast of Ellesmere Island in the Canadian Arctic Archipelago. It is fed by glaciers flowing from the mountainous terrain in the background. Large chunks of such shelves break off and float away within the sea ice, which covers the ocean in the lowermost part of this photo. They compose the famed ice islands, up to 100 meters thick, lying within the sea ice, which is normally only a few meters thick. Such drifting islands have been used as bases for research stations investigating the Arctic Ocean basin. The dark, irregular line (lower left) is a crack (open lead) in the sea ice. (U.S. Geological Survey photo by Austin Post, July 24, 1964.)

crystal structure. Ice is a mineral, although most of us may not think of it as such. It is an unusual mineral because of its low melting point, zero degrees **Celsius** (0° C). Most minerals melt at temperatures of hundreds of degrees Celsius, and we do not usually see them in their molten state, as we do ice. Glacier ice is a **monomineralic** rock, one consisting of a single mineral. Other examples of monomineralic rock are pure **marble**, consisting wholly of **calcite** (CaCO$_3$), and rock salt, composed solely of **halite** (NaCl).

Most glacier ice has had a complex history; it begins as a **sediment** and ends up as a **metamorphic rock**. **Sedimentary rocks** are composed of an aggregate of mineral or rock particles deposited on the earth's surface by any of several processes. The unconsolidated material is sediment; the consolidated state is sedimentary rock. Metamorphic rocks are formed from preexisting rocks by heat, pressure, stress, and recrystallization.

These definitions enable us to recognize that fresh snow is a sediment, which is converted by compaction and solidification into a sedimentary rock, which is metamorphosed by pressure, stress, flow, and recrystallization into glacier ice. Thus, large glaciers are composed principally of metamorphic rock. Their headwater parts contain considerable sediment and sedimentary rock. Very small glaciers may consist largely of sedimentary rock, and all glaciers have a temporary blanket of sediment in winter in the form of new-fallen snow.

A third class of rocks, **igneous**, comprises those formed by the crystallization of minerals from molten substance. Thus, an ice cube in our refrigerator, though an artifact, is like a small chunk of igneous rock. Most geologists do not regard ice formed in nature by freezing of water as an igneous rock, so, respecting their sensibilities, let us call such ice **pseudo-igneous**. Glaciers have **veins**, pods, and individual **crystals** of ice formed directly in place by freezing of water, so they contain a small amount of pseudo-igneous material.

A GLACIER'S BUDGET

Glacier budgets involve items of income and expenditure, and contrary to the practice of many modern institutions, glaciers believe implicitly and wholeheartedly in balanced budgets, which they maintain by advancing or receding. A glacier's income consists almost solely of snow, as we have already recognized. Its expenditures are the product of **wastage** that disposes of snow, or of the ice made from it, primarily by melting (**ablation**) and calving. **Calving** involves the breaking away of large chunks of ice at a glacier's margin. During winter a glacier normally accumulates snow over its entire surface, but in summer this snow disappears from the lower parts of the glacier. If this annual layer of snow melted entirely, we would have a most impoverished glacier on our hands. Let's talk about prosperous glaciers, which retain a good fraction of the annual increment of snow over a significant part of their surface.

Any glacier has two principal parts, an upper or interior area, where more snow accumulates each year than melts, and a lower or outer area, where all the snow melts (Fig. 1.3). In this lower part, some of the glacier ice also melts as it is exposed from beneath the disappearing snow layer. It is reasonable that the upper part, with its excess of snow, is referred to as the **accumulation area** (Fig. 1.4), and the lower part, with a deficiency of snow, as the **wastage area** or **ablation area**.

The boundary between these two areas, as determined by measurements of accumulation and wastage made for an entire year, is what glaciologists term the **equilibrium line** (or **equilibrium zone**, when it is patchy and irregular). Observers looking at a glacier near the end of the melting season see the edge of last winter's snow blanket, the **snowline** (*1* in Fig. 1.5), and possibly the edge of older snow blankets, extending beyond the snowline, which is known as the **firn line** (*5* in Fig. 1.5). If both are seen, the snowline is closer to the equilibrium line than the firn line. **Firn** is simply coarse-grained snow, from one to several years old. In any single year the edge of the annual snow blanket on a glacier at the close of the melting season may lie above, below, or about at the firn line. Making allowances for displacement of the firn line by glacier movement, the relative position of the snowline indicates whether that particular year was favorable, unfavorable, or about average in terms of accumulation, compared to several preceding years. More significant are measurements of the position of the equilibrium line made over five to ten years, which show whether the glacier is stable or in a condition of expansion or contraction.

A glacier is a dynamic, living thing creating a product in one place and, like a commercial business, transporting it to another place where it can be used. After snow gathers in the accumulation area, the glacier transfers it to the wastage area, converting snow and firn to glacier ice in the process. In the wastage area ice can be disposed of by melting, calving, or some other mechanism. When accumulation is unusually large for several years, a glacier expands by advancing so that wastage can be greater. When accumulation is abnormally small, the glacier recedes, decreasing the size of the wastage area and thus reducing melting. In this way it maintains a balance of income and expenditures. Naturally, adjustments do not occur instantaneously; a glacier is a large mass of slow-moving ice. The system has considerable inertia, and time is required

Figure 1.3. This unusual view shows both the accumulation area (A) and wastage area (W) of a small unnamed ice stream (here informally christened Budget Glacier) lying within the St. Elias Mountains of Canada's Yukon Territory, west of the Alsek River. It was photographed late in the exceptionally warm, dry summer of 1951. The sharp line (FL) separating the two areas is the firn line. This was a poor year for glacier nourishment, and somewhat dirty bands of older firn (F) project from beneath cleaner patches of 1951 winter snow (S).

7

Figure 1.4. This spectacular southwesterly view in Canada's Yukon Territory shows a huge (75 km × 35 km) accumulation area supplying ice to the large Seward Glacier, the principal feeder of the Malaspina piedmont sheet (Fig. 2.3), which lies on the coastal plain of Alaska's south shore. The Yukon–Alaska border extends roughly east–west along the skyline ridge through Mount Augusta (4,270 meters), 25 kilometers distant, seen just right of the man, and the pointed right-center peak, Mount St. Elias (5,490 meters). Snow, firn, and ice probably exceeding 1,000 meters in thickness underlie this extensive flat, which with the Malaspina sheet constitutes one of the great glacier systems of North America. (August 1948.)

to accomplish any change. Nonetheless, glaciers are sensitive bodies, able to respond to changes in environment affecting the processes of accumulation and wastage.

PROCESSES OF ACCUMULATION AND WASTAGE

Clearly, the budget is an important item, and it is usually one of the first things an investigator wants to know in studying any particular glacier. Because the processes of accumulation and wastage are the fundamental elements determining the budget, we need to consider how those processes act and how they are related to environmental factors.

Figure 1.5. An unusually dry, warm summer in 1958 made possible this unusual view of the firn-line area of Humes Glacier in the Olympic Mountains of northwestern Washington State. The white area on the left (1) is 1957–58 snow, and its edge defines the snow line. To the right four bands of older, increasingly dirtier firn (2, 3, 4, 5) are exposed. Banded and still dirtier glacier ice (6) lies to the right of the oldest layer of firn. Crevassed glacier ice is visible in the background. (August 21, 1958.)

Accumulation

Snowfall is by far the major source of substance for making glacier ice. Other ways of creating ice, by the refreezing of meltwater or the formation of rime or hoarfrost, are secondary. (**Rime** is granular ice formed from minute water droplets in windblown mist or fog. **Hoarfrost** [Fig. 1.6] consists of delicate skeletal ice crystals formed by the condensation of water vapor.) Because glaciers are nourished primarily by snow, they are strongly influenced by meteorological processes and **parameters**, such as the source of moisture, average temperature, total precipitation, movement of air masses, and winds.

For example, some of the heaviest snowfalls and deepest snow accumulations on record occur in coastal mountains adjacent to open

Figure 1.6. Delicate and beautiful, skeletal hoarfrost crystals on the smooth ice surface of a frozen stream (Cache Creek) in the lowland area north of Mount Denali (McKinley), Alaska, as formed overnight in midfall season. Hoarfrost adds a little to the sustenance of glaciers, but not much.

oceans in cool, temperate environments, which provide large quantities of water vapor, the mountains of Alaska and Scandinavia being good examples. Very cold conditions are not always optimal for glacier development, because cold air holds only minimal amounts of water vapor. Likewise, landlocked interior areas, which at first glance look like settings favorable to glacier formation, may not do well at all, because too much moisture has been wrung out of storms and air masses before they get to those areas. Even regions endowed with heavy snowfall must conserve some of it year after year if they are to have large glaciers.

Once snow has fallen, it can be drifted by wind, and topographic configurations come to play an important role. Large snow drifts can form leeward of protecting ridges and divides and in sheltered amphitheaters. Some glaciers occupying narrow valleys receive significant amounts of snow by way of avalanches from steep valley walls. This might be called displaced accumulation.

Figure 1.7. The highly crevassed terminus of Miles Glacier in Copper River Valley, Alaska, sheds icebergs by calving into a lake. Tidal glaciers (see color section, Plate I) calve bergs directly into the ocean. For such glaciers, calving can be a major mechanism of wastage. Calving also creates and maintains the 50-meter cliff of this glacier terminus. (Photo by Bradford Washburn.)

Wastage

Processes controlling wastage are more subtle and complicated than those affecting accumulation. Many of the world's largest glaciers may be wasted principally by calving, but most glaciers are more extensively affected by melting. Calving (Fig. 1.7) occurs primarily where glaciers enter the sea (Plate I*). The product is icebergs that float away to melt or to threaten ships in lanes of commerce. Calving also occurs into lakes along glacier margins and to some extent, into rivers and streams flowing

* Color plates appear at the middle of this book.

alongside or beneath glaciers. Calving is by far the most important means of wastage for ice shelves floating on the sea (Fig. 1.2), as well as for some ice streams in high latitudes pouring out of ice caps and ice sheets into deep, narrow valleys flooded by the ocean. The amount of calving from an ice margin is influenced by crevassing within the glacier, by the rate of glacier flow, by whether the glacier margin is grounded or afloat, and therefore by the depth of water fronting the ice. Huge tabular icebergs many kilometers across break free from floating ice shelves.

The simple process of melting is influenced by a complex interplay among a host of factors, such as solar **radiation**, cloud cover, cloud thickness and density, reflected and **emitted radiation**, rain, vapor condensation, air temperature, heat conduction, gaseous **convection**, wind, humidity, reflecting power of the snow or ice surface (**albedo**), dust or dirt on the snow, and various dependent interrelationships between these and other factors. Stated most simply, melting is accomplished primarily by solar radiation, secondary radiation from the sky (clouds), and atmospheric heat conduction.

Sunshine is one of the best ways of melting ice or snow, and almost anything that decreases the amount of solar radiation reaching a glacier's surface, for example, a dense cloud cover, reduces the amount of melting. Interestingly, however, sometimes melting by radiation may be greater under a fog or thin cloud cover than under perfectly clear skies, because the fog or clouds return so much outgoing radiation to the glacier's surface by reflection and reradiation.

Rain can melt ice and snow owing to the relatively high specific heat of water. A raindrop only a few degrees warmer than 0° C contains enough heat energy (calories) to melt a small amount of ice. **Specific heat** is the number of calories required to raise the temperature of 1 gram (gm) of a substance by one degree Celsius.

Water vapor condensing on the snow surface is much more effective than rain. Each gram of vapor gives up 539 calories in changing to the liquid state, and it takes only 80 calories to melt a gram of ice.

Wind promotes wastage by constantly disturbing the layer of air directly in contact with the surface. Under windless conditions, a thin layer of air lying against ice becomes chilled to 0° C and loses much of its water vapor through condensation. In that condition it acts as a blanket insulating the glacier from heat within warmer air higher above the surface. Wind disrupts that insulating layer through turbulence. Wind can also transport loose, cold, dry snow from a spot where melting is minimal to a place where melting is more rapid. The best of all worlds, in terms of wastage, is to have warm air heavily laden with water vapor blowing continually across a glacier.

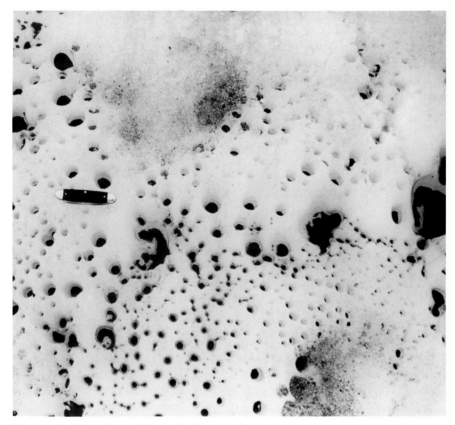

Figure 1.8. The influence of fine dark silt on the ablation of ice is shown by dust wells perforating the surface of Steele Glacier, St. Elias Mountains, Yukon Territory, Canada. The wells are up to 20 cm deep and from 0.5 to 4.0 cm in diameter. The layer of fine dark silt, 1 to 2 mm thick on the bottom, absorbs solar radiation and melts downward until shaded by the walls of the hole. The wells are filled with water.

Surficial **rock debris** can significantly influence melting because of its great absorption of solar radiation. White, pristine snow has a high reflectivity and returns so much radiation to the sky that melting is reduced. A light sprinkling of dark dust or fine dirt lowers reflectivity markedly. Moreover, after a little melting, the fine debris commonly becomes concentrated into small clots that melt downward into the ice, making cylindrical water-filled holes, mostly a few centimeters (cm) in diameter and 10 to 20 cm deep. These **dust wells** tend to form in clusters (Fig. 1.8). The holes look like they were bored by a carpenter's brace and bit, with a millimeter (mm) or two of dark silt added to the bottom. In some circumstances washbasin-like depressions with a similar bottom layer of silt are formed.

By contrast, thicker layers of rock debris act as an insulator, at least partly protecting the underlying ice from melting as rapidly as nearby

Figure 1.9. If you dug into this pile of gravel, you would discover that it is actually a cone of ice mantled by gravel. This is the surface of stagnant lower Steele Glacier, Yukon Territory. The cone is about 5 meters high, and the gravel cover is 20 to 25 cm thick. The cone's symmetry reflects the homogeneity and mobility of the loose, well-sorted gravel mantle. This localized gravel deposit accumulated in what was formerly a depression on the ice surface. The present mound represents an inversion of relief created by insulation of the gravel mantle.

clean or thinly mantled ice. As a result, irregular features of upstanding relief are created, and some may be strikingly symmetrical in shape if the insulating deposit is localized, relatively uniform, and subject to uniform redistribution (Fig. 1.9). If the debris mantle is heterogeneous in distribution and thickness, a chaotic topography of knobs, ridges, and depressions can be created by the contrasting effects of insulation and melting (Fig. 1.10).

An eye-catching feature on some glaciers sparsely mantled by coarse debris are large rock fragments perched on ice pinnacles (Fig. 1.11). These **glacier tables** develop because ice shaded and insulated by the rock melts less rapidly than surrounding unprotected ice. On glaciers in the northern hemisphere tables slowly develop a tilt to the south owing to greater melting on that side from reflected solar radiation. Eventually, as a result of increasing height (usually to a meter or two) and tilt, the rock fragment slides off and leaves a bare ice pinnacle to waste away while the stone starts to form a new table.

Long-continued melting of dead or stagnant ice eventually produces a surface mantle of rocky debris so thick and uniformly distributed that

Figure 1.10. If this scene appears disorganized and chaotic to you, that's the way it should. You are looking down onto the surface of the stagnant lower Steele Glacier, St. Elias Range, Yukon Territory, as it appeared in July 1941. The glacier is about 1.5 km wide here and 75 to 100 meters below you. Ice is seen principally in steep cliffs (I) undercut by ponds or streams. Most of the surface bears a mantle of stones, sand, and gravel only a fraction of a meter thick. A large muddy pond (P) is a little left of center, and four more ponds are visible to the right. Some of the ice cliffs (I) are 15 to 20 meters high. Most of the ice appears dirty because of debris within it and washed down over the face from above. The irregular topography is the result of differential melting under an uneven mantle of debris, dissection by streams (S) running on the ice, and undercutting by ponds. Most glaciers with stagnant lower reaches became over-extended because of short-lived episodes of extremely rapid advance known as surges. The ice surface before stagnation and wasting set in was at the level of the bench on the far side of the glacier. The scene shown here was totally destroyed by a surge of Steele Glacier in 1965–66. For an aerial view of a stagnant glacier surface, see Figure 4.3.

it is stable enough to support vegetative growth (Fig. 1.12). Dense forests with trees exceeding 100 years in age can grow on the surface of such ice bodies (Fig. 1.13). Melting of the underlying ice is so slow that it does not seriously disturb the surface mantle. The forest and the debris mantle can, however, be destroyed locally by slides, by the recession of ice walls enclosing ponds or lakes, or by the cutting of stream channels.

Figure 1.11. The large, perched block of rock has protected the underlying ice pedestal from melting, largely by solar radiation, as rapidly as the surrounding unprotected ice on the surface of Hayes Glacier in the Alaska Range. Southward tilting in northern latitudes (here about 60°30′) is caused by greater reflected-radiation melting on the south side of the ice pedestal. (Photo by Bradford Washburn.)

Figure 1.12. Dark cliff beyond man is a face of stagnant glacier ice (I) discolored by dirt washed down from an overlying mantle of residual ablation detritus on the surface above. A dense cover of vegetation, including trees (T) outlined on the skyline, grows in this mantle atop the glacier. The parachuted loads have been dropped on an outwash sand flat, and the bare bouldery slope, midleft, is ablation debris dumped off the ice. The locality is along the south margin of Malaspina Glacier, a large piedmont ice sheet on Alaska's south coastal plain. (Early July 1949.)

Figure 1.13. Spruce tree about 100 years old, growing in a meter of stabilized ablation debris on stagnant ice near the outer edge of Malaspina Glacier. Melting of ice under the thick debris cover proceeds so slowly and uniformly that trees can take root and survive. Trees like this are shown on skyline of Figure 1.12.

CONVERSION OF SNOW TO GLACIER ICE

Glaciers consist of intergrown crystals of solid ice ranging from the size of a sand grain (Fig. 1.14) to larger than a human head (Fig. 1.15). These crystals have been built from thousands of tiny snowflakes.

The conversion of snow to glacier ice is not a trivial process. It begins almost as soon as snowflakes settle onto the accumulation area, and it is strongly influenced by temperature, proceeding more rapidly in warmer environments. The vapor pressure of water is higher around the sharp

17

Figure 1.14. Crystals within fine-grained cold glacier ice from a borehole at a depth of 70 meters at Byrd Station in western Antarctica, as it appears in a thin section viewed under a polarized light analyzer, which makes some crystals appear dark. Smoothly rounded forms are air bubbles. Faint 1-cm overlay grid gives scale. Crystals in cold ice do not remotely approach the size of those in warm ice (Fig. 1.15). (Photo by A. J. Gow.)

points of the skeletal ice crystals (snowflakes) than over smoother parts, so vapor moves from the sharp projecting points, blunting them and filling in the intervening hollows where it condenses. This **vapor transfer** ultimately produces small, solid grains of recrystallized ice of roughly spherical shape and uniform size. Breakup and compaction of delicate flakes under an increasing overburden of accumulating snow and localized melting at pressure points and refreezing of the liquid in other places aid the process. The resulting loosely packed mass of small uniform ice grains, known as old or granular snow, is termed **névé** by the French or **firn** by the Germans.

Firn is officially defined as snow that is at least one year old, having survived an episode of summer melting. It generally has the physical

Figure 1.15. The aggregate of coarse ice crystals, some as big as your head, in this face of stagnant Malaspina Glacier ice started as hundreds of thousands of delicate snowflakes in the St. Elias Range 100 km to the north. A muddy stream undercuts the base of the cliff, and the branch of an alder bush with leaves up to 7 or 8 cm across leans against the ice.

characteristics of granularity, permeability, and increased density compared to cold, fluffy, new-fallen snow, which may have a density as low as 0.06. (**Density** is the **mass** of a substance per unit volume. Liquid water has a density of 1.0; that is, a cubic centimeter weighs 1 gm.) Grain size and density increase with age in firn, through action of the processes just described, and old firn commonly has densities ranging from 0.4 to 0.8, depending strongly, among other things, upon the thermal environment. Ice grains in low-density firn are loosely bound, with considerable open space between, so the assemblage is porous and permeable, but both properties decrease with increasing density. At densities between 0.80 and 0.85, the grains become more tightly bound, the remaining open space is reduced, and, most importantly, the openings between grains become sealed, so the mass has limited permeability (Fig. 1.15). This is the state of glacier ice.

This is also close to the time at which the glacier starts to undergo solid flowage. **Solid flow** involves recrystallization of the ice, which eliminates many of the remaining air bubbles and increases its density to 0.90; the maximum density of pure, bubble-free ice under atmospheric pressure at sea level is 0.92. There are several denser, high-pressure forms of ice, but they do not exist in glaciers.

The largest and clearest ice crystals in glaciers usually exist in nearly inactive or totally stagnant ice, which has recrystallized in situ. The process of solid flow, accompanied by repeated recrystallization, keeps the crystals in rapidly flowing glaciers more modest in size. Walking across the surface of a clean glacier consisting of crystals 10 to 20 cm in diameter on a warm or rainy day, when the surface is wet, can be something of an adventure. It is akin to walking across a field of loose ball bearings of similar size. The rounded surfaces of the ice crystals are so slick that crawling becomes the only reasonable mode of locomotion.

An interesting and beautiful aspect of large, clear crystals of glacier ice is found in the **melt figures** that develop within them after 15 to 30 minutes of exposure to bright sunlight. These are thin, circular, plane disks, commonly 4 to 6 mm in diameter, containing water and a bubble of vapor. Total reflection of light occurs from them under proper orientation, making a glistening, diamondlike display. Within any single crystal they are all parallel, lying in the basal crystallographic planes. First described by famed British physicist and alpinist John Tyndall in the middle of the nineteenth century, they are often referred to as **Tyndall figures**.

THE COLOR OF GLACIAL ICE

The color of any object viewed by human eyes is determined by the quality (**wavelengths**) of light that is reflected or emitted by an object or has passed through it. Daylight appears colorless, although it consists of a continuum of different wavelengths that, if separated, are perceived by the eye as ranging from red to blue. This is shown by the **refraction** of light through a triangular prism of transparent material, such as glass, which breaks the light into a **spectrum** of colors. Light travels as a series of **electromagnetic** waves that carry small bundles of energy known as **photons**. The red end of the visible spectrum has a wavelength of about 700 nanometers (nm)* and the blue end about 400 nm.

Many travelers in polar or alpine regions have noted the bluish color displayed by large chunks of glacier ice – icebergs, for example. It can range from a faint tinge to a gorgeous, brilliant, deep shade of clearest blue. Blue is also seen in deep holes or caves in snow. Not all glacier ice is blue, for much appears to be white. The whiteness is caused by the strong reflection of incident white light from surfaces between air and ice created by the abundant bubbles trapped within such glacier ice and by reflection from the rough surface of bubbly ice. The foam on many liquids, beer for one, is white for the same reason, even though the liquid may be colored. This strong reflection easily overrides the faint bluish color of ice. The eye sees ice as bluish if the crystals are clear, devoid of air bubbles, and tightly grown together. A mass of such ice several meters in size is usually required to produce a perceptible coloration.

A frequently heard question is, "What causes the blue color of ice?" Because light reflected from ice is unquestionably white, the bluish color must be the result of differential absorption of waves transmitted through ice or of waves secondarily emitted under excitement by impinging daylight, or both. Historically, emission resulting from **molecular** (or **Rayleigh**) **scattering** has been assigned a prominent role, but modern opinion based on accurate measurements strongly favors **absorption** of red wavelengths in transmitted light as the primary cause of the blue color, with emission playing a minor role at best.

Emission resulting from molecular scattering, the accepted explanation for our blue sky, was first analyzed by the famed British physicist, Lord Rayleigh, in the early 1870s. It occurs as follows. Substances consist of **atoms**, and every atom has a core of particles, **neutrons** and positively charged **protons**, surrounded by orbiting **electrons**, each of which carries a negative electrical charge. These electrons can be set into oscillation by the electromagnetic waves and photons of incident light. This oscillation causes the electrons to emit light outward in all directions; hence the term *scattering*. The light emitted from ice crystals by molecular scattering is known to be enriched in blue wavelengths compared to

---

* A nanometer is one-billionth of a meter, or 0.000001 mm.

incident daylight, but the effect is weak and by itself probably not perceptible to the human eye.

Recent accurate meaurements of absorption of light waves transmitted through clear ice has shown that absorption at the red end of the spectrum is about six times greater than at the blue end. This occurs because the incident light sets the ice **molecules** into vibrations that consume some of the light's energy. Both red and blue light participate in this activity, but the lower **frequency** (number of waves passing a point per second) of the red light makes it more effective in generating vibrations and more of it is consumed, thus absorbed. The human eye perceives the diminution in red as an enhancement in blue.

Although the water molecule is structurally simple, consisting of one oxygen atom and two hydrogen atoms, water is anything but a simple substance with respect to its chemical and physical properties. The molecules in ice, and scattered aggregates of molecules in water, are held together by an unusual force known as **hydrogen bonding**. Think of the molecules as held together by rubber bands that can stretch and contract. This is a strong bond, and considerable energy is required to cause the molecules to vibrate. For that reason ice produces a stronger blue light by absorption of red than many other substances. The hydrogen bond also accounts for other unusual characteristics of water, such as its high boiling point and the fact that the solid state is less dense than the liquid. Ice floats on its own liquid unlike most other solid substances.

In a sentence: Ice appears blue primarily because daylight transmitted through it is preferentially absorbed at the red end of the visible spectrum, producing an impression of the color blue in the human eye.

RECAPITULATION

Glaciers are a special kind of rock composed of the mineral ice, which initially accumulates on land as snow. Glaciers convert soft, fluffy snow into solid, coarse-grained ice and transport it from an area of accumulation to an area of wastage. There it is disposed of primarily by melting and calving. These bodies are dynamic, highly sensitive to the climatic environment and especially to variations that influence the processes of accumulation, flow, and wastage. The following chapter examines the interesting features and relationships that flow creates within glaciers, as well as the types of glaciers that characterize different settings and environments. We have just started to penetrate the intriguing and mysterious realm of glaciers; let us press on.

2

Types, features, and characteristics of glaciers

As we prepare to create our glacier, we need to consider the variety of glaciers that exist and the particular type we propose to create.

KINDS OF GLACIERS

A number of highly detailed classifications have been proposed, but glaciers can be most easily differentiated on the basis of geometry and internal temperature regime.

Geometry of glaciers

Geometrically, glacier ice composes streams, sheets, caps, or carapaces. **Ice streams** are exactly what the words say, and many occupy the canyons and valleys of mountainous terrains. They start in basins near mountain crests and flow down the canyons. The term *ice streams* is also used for narrow linear zones of especially fast-flowing ice within large ice sheets, although the words *ice current* might be more appropriate. **Ice sheets** usually occupy broad areas of gentle **terrain**, but they can become so thick as to overwhelm the underlying terrain, regardless of relief. Huge sheets generally flow outward in all directions from a central region or from subsidiary centers of high accumulation. Flow in special directions may, however, be guided by local subglacial topographic configurations, especially under thin marginal ice. Sheets of ice can be of continental proportions, for example, the Pleistocene Ice-Age sheets of North America and Eurasia, which possibly attained respective thicknesses of nearly 5,000 and more than 3,000 meters. (The **Pleistocene Epoch** embraces the past 1.6 million years of geological time.) The ice of Antarctica is a present-day example of a continental sheet, with a maximum thickness

Figure 2.1. This relatively thin carapace of ice cloaking the north-northeast side of Mount Rainier in Washington State feeds simple outflowing ice streams. Pronounced crevassing results from rapid flow down the steep slopes of this volcanic cone. (U.S. Geological Survey photo by Austin Post, September 11, 1964.)

approaching 4,776 meters, as determined by several types of reflected geophysical soundings, including classical seismic procedures and airborne radar. The subcontinent of Greenland harbors a much smaller sheet, but its maximum thickness of 3,100 meters indicates that it is not a trivial body of ice.

Smaller sheets form **ice caps** on upland flats, such as plateaus, and some completely submerge highlands with mountainous relief. **Ice carapaces** commonly cloak high, conical volcanic peaks (Fig. 2.1). Any sheet of ice or ice carapace may feed ice streams flowing down valleys extending outward from its margins. These are **outlet glaciers**, and some of them

Figure 2.2. Small piedmont bulb of ice near Dobbin Bay on Ellesmere Island, Canadian Arctic. Ice tends to spread into bulbs like this whenever a steeply descending ice stream debouches vigorously onto nearly flat terrain. Radial pattern of crevasses results from circumferential extension. When and if this glacier recedes, a piedmont lake will probably be impounded inside the morainal accumulation (M) that has been built around the ice edge. (U.S. Geological Survey photo by Austin Post, July 19, 1964.)

are huge, carrying great discharges of ice supplied by large parent bodies. Jacobshavn Glacier flowing into Disko Bay on the west coast of Greenland is a good example. Glaciers like it pour thousands of icebergs into the Atlantic Ocean every year. Some arctic and antarctic outlet glaciers upon reaching the sea spread out upon its surface as a floating shelf of ice. An **ice shelf** is simply a special type of ice sheet. The Ross Ice Shelf of Antarctica is a large, much studied example, but there are many others in the Antarctic as well as a few in the Arctic, one being off Ellesmere Island (Fig. 1.2). If an ice stream from a mountainous area or from a local ice sheet debouches onto flat land rather than into the sea, it may spread out into a broad lobate sheet known as a **piedmont** (foot of the mountain) **glacier** (Figs. 2.2 and 2.3).

Figure 2.3. The Malaspina is probably North America's premier piedmont glacier, a sheet of ice covering an area of over 5,000 km² on the flat coastal plain at the foot of the lofty St. Elias Range in southern Alaska. This northwestward view is along the eastern edge of the Seward Lobe of the Malaspina. The dark chevrons are highly attenuated medial moraines of ice streams from the St. Elias Range, possibly deformed by surging of the feeding ice stream or by crumpling as the ice spreads into a sheetlike bulb. Ice in the left foreground is nearly 600 meters thick. Mount St. Elias (5,490 meters) is the sharp peak in the background. (Photo by U.S. Coast Guard, August 1953.)

An ice stream in a mountain valley is the type of glacier we are most likely to see, so let us bend our efforts in that direction and devote our attention in succeeding pages mostly to ice streams.

Warm glaciers and cold glaciers

We have a choice of making a warm or a cold glacier. This may sound a bit silly, because all glaciers are relatively cold; nonetheless, important differences exist in degrees of coldness. Some writers avoid this semantic problem by employing the terms *temperate* and *polar*. Let us examine the considerations involved.

Pure water at sea-level atmospheric pressure freezes at 0° C. At pressures significantly greater, temperatures a little lower are required to produce freezing. Increasing pressure lowers the freezing point of water at a rate of 0.0072° C per atmosphere.* This means that at pressures greater than 1 atmosphere, temperatures must drop below 0° C before pure water will freeze; conversely, **pressure melting** of ice at 0° C occurs under pressures exceeding 1 atmosphere. One reason ice skates glide so easily is that the concentrated pressure under the narrow steel blade is great enough to cause local melting, and the water lubricates passage of the skate blade. After pressure is released, the meltwater refreezes.

At the base of a glacier 1,500 meters thick, the pressure is great enough to lower the melting point by one degree below 0° C (Fig. 2.4A). Thus, within this 1,500-meter-thick glacier the ice must become colder and colder with depth in order to escape melting. A warm glacier must have some mechanism for cooling itself internally. It cannot do so by simple conduction to the surface because the temperature gradient is in the wrong direction (Fig. 2.4A). Actually, the mechanism is simplicity itself: As the glacier grows thicker through accumulation on the surface, a little ice within the glacier melts because of the increased pressure. The heat required for melting, 80 **calories** for each gram of ice, can come only from the body of the glacier itself, resulting in the cooling of the remaining ice.

A glacier at the **pressure-melting temperature** of ice throughout is by definition a **warm** or **temperate glacier**. A layer of ice a few meters thick at the surface can be temporarily chilled below freezing in winter, but this chilled layer is rapidly brought back to the pressure-melting temperature next summer. It has no permanent effect on the glacier's internal temperature regime. Glacier ice at the melting point is in equilibrium with liquid water, so water can exist throughout warm glaciers all the way to the base.

In contrast, a **cold** or **polar glacier** has a temperature below the melting point of ice from top to bottom. Such glaciers get warmer, rather than colder, with depth (Fig. 2.4B). Curiously, this is owing to the fact that temperatures within Earth increase with depth. The temperature in some deep mines is so high as to be virtually unbearable for humans – on the order of 55° to 60° C (131° to 140° Fahrenheit [F]). This is owing to the fact that heat is continually being conducted from within

---

* One **atmosphere** is defined as a pressure of 1 million dynes per square centimeter (cm^2) at sea level. A **dyne** is a unit of force capable of accelerating the movement of 1 gram of **mass** by 1 cm per second for every second it is applied.

RELATIVE TEMPERATURES

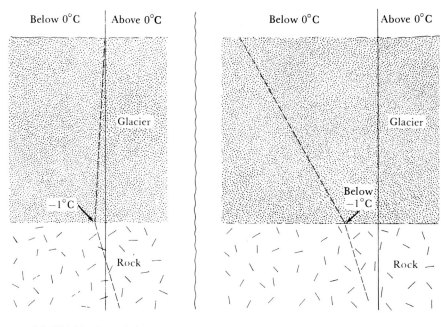

Figure 2.4. Cross-section sketch of thermal relationships within a warm and a cold glacier. Dashed lines represent temperature gradients. In summer a warm glacier is colder at its base than at the surface because pressure exerted by overlying ice lowers the freezing point. In order to qualify as cold – that is, to be below the melting point throughout – a glacier must have a surface temperature low enough to prevent melting at the bottom by Earth's internal heat, which has to be conducted through the ice to the surface of the glacier. Liquid water can exist within and at the base of a warm glacier, which is at the pressure melting temperature of ice from top to bottom, but a cold glacier cannot contain liquid water anywhere except temporarily under special circumstances. Warm glaciers melt at the bottom; cold glaciers do not.

Earth to its surface at an average rate of about 38 calories per cm² per year, the so-called **geothermal flux**. Because a cold glacier is warmer at the base than at the surface, it can conduct the geothermal heat flux to the glacier's surface, where it is dispersed into the atmosphere. To do this, the surface of a cold glacier, 1,500 meters thick, has to have a mean annual temperature of about −40° C, so cold glaciers exist only in truly frigid environments. Because they contain essentially no liquid water internally or at the base, they behave differently from warm glaciers. Heat arriving at the base of a warm glacier cannot be conducted to the surface because the temperature gradient within the ice is reversed, colder at the bottom than at the top (Fig. 2.4A). As a result a little ice is melted at the bottom of warm glaciers by the geothermal flux, about half a centimeter each year.

This representation of thermal relationships within glaciers is simplified, but in principle it is sound and serves our purposes. A large glacier need not be totally warm or cold in all its parts. Where thickest, both the Antarctic and Greenland ice sheets are cold toward the surface and warm near the base, but in thinner outer parts they are cold from top to bottom. Even presumably warm glaciers may display unexpected thermal anomalies. Warm glaciers are more numerous, more active, and more interesting, so our discussions will focus principally on them.

THE MAKEUP OF VALLEY GLACIERS

Valley glaciers differ considerably from one another, and some are highly complex. In its simplest form such a glacier is a stream of ice flowing down a valley from an accumulation area at its head, usually located in a theaterlike basin or assemblage of basins. Unlike a stream of water, it flows slowly, perhaps a fraction of a meter a day, and it may be as much as 1,000 meters deep.

Any stream of ice moving down a valley picks up dirt and rock along its edges, forming a marginal zone of dirty ice. Below the firn limit, where melting dominates, wastage of this dirty ice causes debris to accumulate residually on the ice forming a linear, debris-covered band called a **lateral moraine**. The debris eventually becomes thick enough to protect the underlying ice, so a ridge of ice mantled by rock debris is formed (Fig. 2.5). This type of lateral moraine on the ice is not to be confused with the depositional lateral moraines left by a receding glacier (described in Chapter 8). When two ice streams join, either as near-equal branches (frontispiece) or with one as a subsidiary tributary, the adjacent lateral zones of dirty ice and their overlying lateral moraines coalesce into a **medial moraine**. Most valley glaciers, like water streams, have tributaries and consequently medial moraines.

An important thing to remember about moraines on a glacier's surface is that they represent the outcropping traces of septa of dirty ice normally extending from the surface to the floor (Fig. 2.5). As a glacier with moraines undergoes progressive melting descending its valley, these morainal ridges get higher and broader because the amount of rock debris accumulating residually on the surface is steadily increasing, thereby providing greater insulation to the underlying ice. It also spreads laterally to form a wider band.

Barnard Glacier of Alaska (frontispiece) is a classic **compound valley**

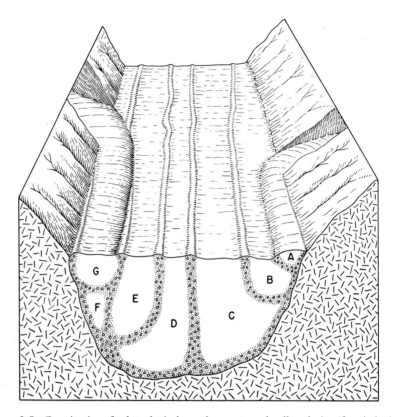

Figure 2.5. Constitution of a hypothetical complex compound valley glacier (frontispiece) representing possible relationships between different types of ice streams. Ice stream A is superimposed on B, which in turn is inset into C, which is juxtaposed side by side with D. E and F are juxtaposed and inset into D, whereas G, initially superimposed on F, has subsequently sunk into an inset relationship. The terminus of such an inset ice stream will be marked by a morainal loop somewhere short of the terminus of the trunk glacier (see Fig. 2.6). When looking at a compound glacier one should always think about what is under the surface as well as what is on it.

glacier made up of more than a dozen separate ice streams, as shown by the number of parallel medial morainal stripes on its surface. Formation of a typical medial moraine, between two principal branches of the trunk glacier, is well illustrated in the middistance up-glacier. Each of three tributaries coming in from the right also forms a medial moraine. The nearest tributary is a compound glacier in its own right, with at least two tributaries indicated by medial moraines. Note that the tributary ice streams make room for themselves by forcing the trunk ice stream to move over and by compressing adjacent ice streams into narrower bands. Although not visible in this picture (frontispiece), those ice streams increase in thickness as they are compressed.

If the bedrock floors of joining glaciers are accordant, that is, at the same level, **juxtaposed ice streams** lying side by side are formed (*C* and

Figure 2.6. This view southwestward to the double summit of North America's highest peak, Mt. Denali (McKinley) at 6,189 meters, also covers the upper part of one of Alaska's more interesting valley glaciers, the Muldrow. The complex of foreground moraines on the ice moved abruptly 6 km down-valley during a surge in 1956–57, and the previously stagnant lower reach was rejuvenated. Moraine A along the right edge of the glacier curves and joins the right bank (looking upstream) in a way that suggests the enclosed ice stream (AS) is an inset glacier (Fig. 2.5). Moraines B and C likewise appear to curve toward the margin, suggesting that ice streams BS and CS may also be inset. If so, the Muldrow may be one of the few glaciers in captivity with three stacked inset streams. (Photo by Bradford Washburn.)

D, Fig. 2.5). Each stream extends from the surface to the floor, and they are separated by septa of dirty ice. If the valley of a tributary glacier has a floor at a higher level than the trunk-valley floor, that ice stream does not extend to the bed of the trunk glacier, but rather occupies an inset position within it (*B* and *G*, Fig. 2.5).

Below the firn limit all glaciers become thinner toward the snout owing to melting. Because any **inset ice stream** is necessarily thinner than the trunk glacier, it is eliminated by melting before reaching the trunk's terminus. As a result, its inner medial moraine curves around and joins the outer medial or lateral, making a convexly curved end moraine on the glacier's surface (Fig. 2.6). This end moraine is composed of the

31

detritus picked up and carried along by the bottom ice of the inset stream while it was flowing down the tributary valley.

The floors of **discordant tributaries** may actually hang above the surface of their trunk glacier, so the tributary descends by means of an **icefall**. It comes to lie on the surface of the trunk glacier in a superposed position (*A*, Fig. 2.5). The weight of a **superposed ice stream** is usually great enough, however, that it eventually sinks into an inset relationship down-valley (*G*, Fig. 2.5). Not all glaciers are compound; some draining from one valley-head amphitheater or forming an outlet from an ice sheet or carapace can be composed of single ice streams.

STRUCTURES WITHIN GLACIERS

Because glacier ice is formed out of a sedimentary accumulation, it is not surprising that traces of sedimentary layering (Plate II), especially the annual dirty bands, are weakly preserved in the ice tongues of some valley glaciers. More prevalent and more typical, however, is another less regular lamination in the ice, known as **foliation**, a metamorphic structure secondarily created by conditions and processes associated with flowing ice.

Foliation

Foliation is expressed in ice by differences in grain size or crystallinity or by air bubbles and dirt content of associated bands of ice (Fig. 2.7). Commonly, the differential response of these bands to surface melting emphasizes the existence and orientation of foliation within glaciers. It is usually strongly developed near the lateral margins of valley glaciers owing to strong stresses between moving ice and immobile valley walls. Marginal foliation is normally steeply inclined (Fig. 2.8), and its orientation is longitudinal, parallel to the flow direction. Zones of foliation of similar character and orientation can be formed farther out in valley glaciers by the differential stress between ice streams from tributary valleys joining at different flow velocities. For reasons not fully understood, these stresses cause recrystallization to occur more completely in some planes than in others, producing differences in air-bubble content and crystal size. This movement also strings out concentrations of rock debris.

We would expect ice near the base of a glacier also to be strongly foliated because of stresses created by movement over the bed. Basal

Figure 2.7. A prevailing structure in glaciers is the irregular, lenticular banding known as foliation. This is what it looks like. In this specimen, 30 cm wide, from Blue Glacier in the Olympic Mountains of Washington, the dark bands are clear, coarsely crystalline ice, and the more whitish layers are finer-grained, granular, and bubble-rich ice. It is this structure that causes geologists to regard glacier ice as a metamorphic rock. It can be created by shearing and by intense compression. (Photo by Barclay Kamb.)

foliation, however, becomes visible only near the glacier's snout. It is usually gently inclined upstream with a trend transverse to the long dimension of the ice stream rather than parallel to it. Experienced travelers, caught in a white-out fog on a glacier, know that they can use foliation as a navigational guide for finding their way home.

Strong transverse foliation can also be developed below icefalls by the intense compression occurring there. This foliation is formed more by compression of a heterogeneous jumble of snow, ice blocks, and rock debris than by shearing. A pure, homogeneous mass can be severely squeezed and stressed without necessarily creating foliation, because there are no differences in material to be smeared out into separate

33

Figure 2.8. Strong near-vertical foliation banding in sheared ice, emphasized by differential surface melting, near the margin of lower Blue Glacier, Mount Olympus, Washington. Alternate bands involve coarse-bubbly, coarse-clear, granular, and fine-grained ice, both clear and bubbly.

laminae. A trash dump composed of old car bodies, tin cans, lawn mowings, woody plants, soft-drink bottles, and so on, indiscriminately jumbled together, is a heterogeneous mess, but it would make a beautifully foliated mass, if severely compressed, for just that reason. The jumble of material at the base of an icefall is a natural trash dump, well suited for the formation of compression foliation.

Ice pouring over some icefalls may be divided into separate streams by bedrock ribs or bastions (Fig. 2.9) projecting from the fall's face and extending well up from its base. These ribs supply rock debris to the

Figure 2.9. Background for this scene is provided by the north (right) half of the 300-meter-high icefall of Blue Glacier, split by a rocky bastion (B). The sharp skyline peak is Mt. Olympus. Icefalls, besides being scenic, create structures such as foliation, longitudinal septa, and ogives within glaciers. Flow velocity accelerates greatly over falls, and the glacier is literally pulled apart, forming many crevasses and becoming so much thinner that some of the deeper cracks penetrate to the bed. Slowing of velocity at the bottom of the fall produces strong compression and greater thickness. Foreground apparatus is used to study size, orientation, and interrelationships of crystals in glacier ice. (Summer 1957.)

margins of the separated ice streams, and when they are reunited at the base of the fall, transverse compression and differential shear of this dirty ice generates a strongly foliated band of ice. This foliation is steeply inclined and longitudinally oriented, and the prominently structured band is carried down-valley within the ice stream forming a feature called a **longitudinal septum**. One that has been mapped in detail within Blue

35

Glacier on the northeast flank of Mount Olympus in the State of Washington was, for reasons too lengthy to detail here, jocularly named the *Gesundheitstrasse* (Fig. 2.10). The ability of icefalls to create structures within glaciers caused some early glaciologists to refer to them as "structural mills."

Ogives

Striking features created by some icefalls are symmetrical, curved, periodically repetitious bands, or surface swells and swales, called **ogives** (Figs. 2.11 to 2.13). Their curvature simply reflects the greater velocity of the central part of the glacier compared to the margins, as explained in Chapter 3, but other aspects of their origin are perplexing, in spite of much study. Dark and light **banded ogives** (Fig. 2.11) are the surface exposure of spoon-shaped, three-dimensional structures inclined upstream within the glacier. The dark bands consist of highly foliated ice, some fine-grained, some coarse-grained, both rich and poor in air bubbles, and usually somewhat dirty. They look like ice that has been broken up, mixed with snow and dirt, and then severely compressed. The light bands are more homogeneous and less strongly foliated and contain ice rich in air bubbles.

Some investigators have proposed that a pair of ogive bands, light and dark, is of annual origin. The light band is regarded as the ice that passed from the icefall into the zone of intense compression below during winter, when snow covered everything and filled crevasses. Dark bands are presumed to represent ice that underwent compression during summer, when open crevasses provided sites for the accumulation of broken fragments of ice and dirty debris rather than snow.

The surficial **swell-and-swale** (bump-and-hollow) type of ogive may indeed be an annual feature. This interpretation is supported by relationships shown in Figure 2.12, where photos show that a new swell-and-swale pair formed in one year at the base of an icefall. It is hypothesized that the segment of ice moving through an icefall in summer is thinned owing to melting, so that it makes a swale when compressed at the base of the fall, compared to the thicker segment of unmelted ice that moved through the icefall in winter and also received a cover of snow.

The relief of wave ogives becomes less marked down-glacier because of differential surface melting, the swales being partly protected by deeper infillings of winter snow. Eventually the swell–swale ogives disappear. In some glaciers, banded ogives are seen to succeed wave ogives down-

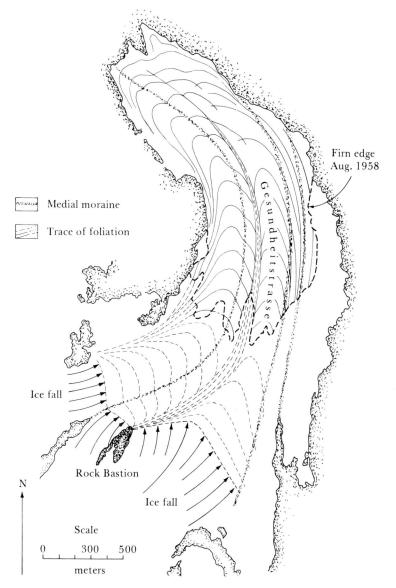

Firn edge
Aug. 1958

Medial moraine

Trace of foliation

Gesundheitstrasse

Ice fall

Rock Bastion

Ice fall

N

Scale

0   300   500

meters

Figure 2.10. This simplified structural map of Blue Glacier shows features produced by passage over the icefall of Figure 2.9, especially the longitudinal septum in the middle, jocularly called the Gesundheitstrasse, *and the pattern of transverse foliation extending to the snout. Icefalls have been called the structural mills of glaciers because of the features they create, which are carried into the ice tongue below the fall.*

valley, but they too commonly disappear before reaching the snout (Fig. 2.11), indicating that the ogive structure usually does not extend to the bottom of the ice.

Figure 2.11. One of the most striking structural features of glaciers are the curved forms known as banded ogives so clearly displayed in the ice tongue of Patmore Glacier, in the coastal mountains of British Columbia. They are created within and at the base of the icefall (F) descending from the large upland accumulation basin (A). The dark bands consist of strongly foliated, recrystallized, dirty ice, the white bands of clean, weakly foliated, bubble-rich ice. They extend almost to the base of the glacier, as shown by the fact that they continue almost to its snout. Many investigators think a pair of bands may be of annual origin, the dark band representing summer and the white band winter. Ogives form only in glaciers with icefalls, but not all icefalls make ogives. (U.S. Geological Survey photo by Austin Post, August 26, 1963.)

Crevasses

The most obvious, abundant, and characteristic structures of glaciers are **crevasses**. These cracks are opened in the brittle crust by flow of the underlying ice. Most crevasses are linear to gently curvilinear, near-

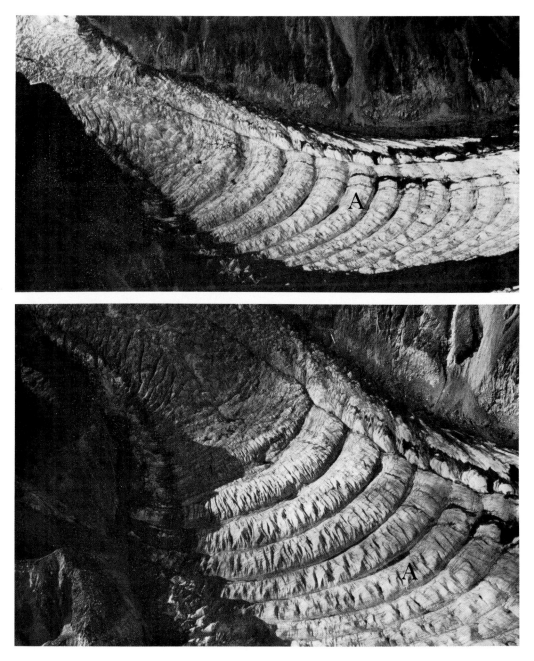

Figure 2.12. A second type of curved conformable features formed by some icefalls is shown by these wave ogives consisting of swells and swales on the surface of Gerstle Glacier in the Alaska Range. By counting upstream from the swell labeled A on these two photos, taken one year apart, one sees that a new swell-and-swale pair was formed in that year, suggesting that the ogives are indeed of annual origin. On some glaciers, wave ogives give way downstream to banded ogives, reinforcing the idea that they too may be of annual origin. (U.S. Geological Survey photos by Austin Post, August 22, 1960, and August 10, 1961.)

Figure 2.13. Alaska's Yentna, a glacier for all occasions. This Austin Post photo, looking northeast in the Alaska Range west of Mt. Denali (McKinley), provides enough information for a one-hour glaciological lecture. The dark longitudinal stripes are medial moraines; the concentric bands of the second tributary on the left are wave ogives. Banded ogives are faintly visible in one of the ice streams composing the right branch of Yentna Glacier (upper center). Both types of ogives are formed in ice falls upstream. Powerful ice streams with high discharge rates make room for themselves by unmercifully squeezing other ice streams (left midfield). Ice is agile; note the abrupt turns executed by tributaries on the right. (U.S. Geological Survey photo by Austin Post, August 31, 1967.)

Figure 2.14. This pattern of domino blocks has been created by crevassing in a fast-moving, unnamed glacier between the Nimrod and Byrd glaciers flowing east out of the Transantarctic Mountains in Antarctica. Long, near-equally spaced transverse crevasses, extending left to right, were formed in a steep reach somewhere upstream, and the short, sharp-edged, orthogonal crevasses developed later, as the ice stream spread out upon emerging from its valley. The rounded edges of the transverse crevasses are produced by wind erosion and snow accumulation, and demonstrate that the transverse set antedates the perpendicular set. (Photo by B.K. Lucchitta, U.S. Geological Survey, December 6, 1984.)

vertical, a few tens to thousands of meters long, and up to many meters wide (Plate III). In warm valley glaciers few are much deeper than 30 meters, but crevasses in cold glaciers can be much wider and deeper. Some antarctic crevasses (Fig. 2.14) are big enough to swallow large tractors – and they have done so. All crevasses are hazardous, and those bridged or hidden by snow are death traps. Some snow bridges (Plate II) can bear traffic, but all are untrustworthy and should be avoided.

Our valley glacier is almost certain to have short crevasses along its lateral margins (Figs. 2.15 and 2.16), pointing up-glacier. These are formed by stretching produced by the large difference in flow velocity

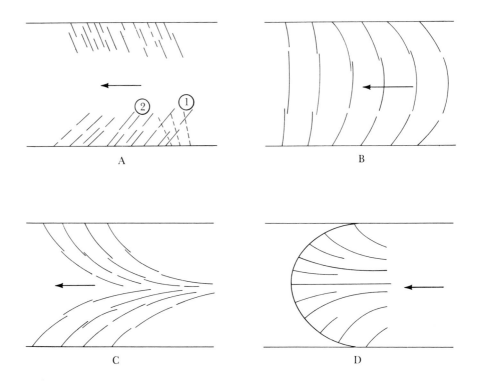

Figure 2.15. Common types of crevasses in valley glaciers: (A) marginal, (B) transverse, (C) splaying, and (D) terminal radial splaying. All such crevasses are the product of orthogonal extension. Dashed marginal crevasses numbered 1 in (A) are older. They formed pointing upstream like 2 but were rotated to point downstream by more rapid flow of ice in the center of the glacier, which also straightened transverse crevasses initially strongly concave downstream (B).

along the glacier's margin owing to the drag of the valley walls (Fig. 2.15A). Almost any part of any glacier experiencing acceleration is likely to develop crevasses transverse to the direction of flow (Fig. 2.15B). Such crevasses may be initially concave down-glacier, but they gradually become straighter (Fig. 2.15B), because the center of the ice stream is flowing faster than the margins. Icefalls are intensely crevassed by the greatly accelerated rate of flow (Fig. 2.9).

In places where an ice stream spreads out, a radially splayed pattern of crevasses may form (Fig. 2.15C and D), and almost any positive relief feature, such as a major bump or ridge on the floor of a glacier, is likely to be marked by one or more sets of crevasses created by doming and extension of the brittle crust as it passes over the bump (*TTC* in Fig. 2.16). Open crevasses can close if they pass into an area of compressing flow, and the position and orientation of all crevasses are changed by flow within the glacier. For example, marginal crevasses that initially point

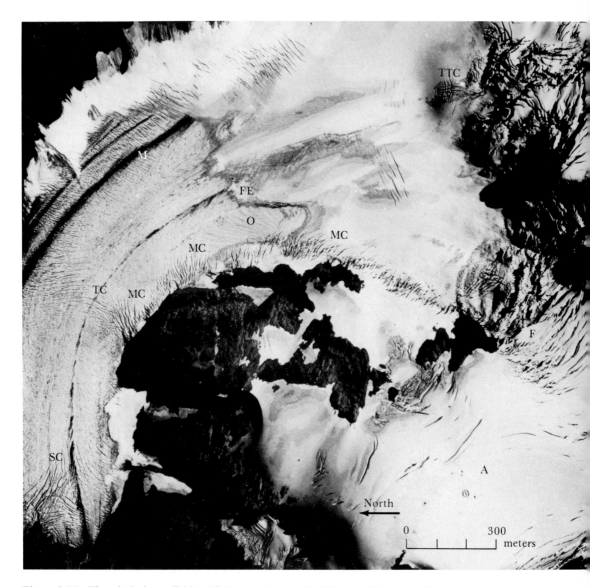

Figure 2.16. The relatively small Blue Glacier on the east side of Mount Olympus in Washington State has served for many years as an outdoor laboratory of glaciological research. This vertical aerial photo shows its snout (S), ice fall (F), and part of the accumulation area (A). Types of crevasses easily identified are splaying (SC) near the terminus, transverse (TC) near the bend, marginal (MC) along the edge, and a tic-tac-toe pattern (TTC) over a dome, upper right. Other features include a faint ogive banding (O), the firn edge (FE), and medial moraines (M). North is to the left. (September 11, 1957.)

up-glacier can be rotated by the faster central flow until they point downstream.

Crevasses allow glaciologists to look into the crust of a glacier (Fig. 2.17), but it is no pleasure to work in them. They are dangerous to

43

Figure 2.17. What is it like to be down inside a glacier? The interior of this fresh, clean-walled, crevasse, less than a meter wide, in upper Seward Glacier, Yukon Territory, gives a partial answer. The darker horizontal bands in the walls are layers of solid ice formed by refreezing of meltwater within finer-grained ice and firn. Irregular projections are hoarfrost. Observer is about 12 meters below the surface. Such a crevasse is not the most comfortable place in which to study glaciers on a warm day owing to a constant rain of ice water falling from above.

approach and hard to enter, generate internal showers of ice water, and inspire feelings of claustrophobia. Dangling on the end of a rope inside a crevasse, a scientist can only hope the glacier will not suddenly close the opening.

MELTWATER ASSOCIATED WITH GLACIERS

We have chosen to make a warm glacier, so it inhabits a reasonably temperate environment. Were we to visit our glacier during summer, we would be impressed by the large amount of meltwater associated with the ice tongue below the firn limit. This is the melting season during which not only the last winter's snow but also some of the underlying ice must be melted and the resulting water carried away.

Melting of snow also occurs above the firn limit in summer, but there the firn is so porous and pervious, in its upper layer at least, that most of the water disappears by percolation. When abundant enough, the percolated water can completely fill all openings between ice grains in the firn, forming what geologists call a **perched water table**. The **water table** is the surface defining the upper limit of water beneath the surface that fills all pores and other openings. In this case, the table is said to be perched because it lies on top of impervious firn or ice, which holds it above the normal water-table level. It is not uncommon upon looking into a crevasse, in such a situation, to see standing water just a few meters down. Exploring such crevasses requires a seal or scuba diver. Near the firn limit where only a thin layer of firn or snow covers impervious glacier ice, concentrations of meltwater create snow swamps. Wading through them can be messy and traumatic, because at any moment one might step into a deep hole filled with ice water.

On bare ice, however, lovely little clear blue pools of meltwater make handy drinking spots for parched travelers. They also look inviting to sweaty hikers for a quick dip, but one must be prepared for a shock. About thirty seconds in water at near 0° C is sufficient. Water overflowing from these pools and derived from melting of the surrounding ice surface gathers into little streamlets that combine to make larger streams, which can carve steep-walled channels into the glacier's surface, tens of meters deep. Some intrenched channels develop meandering courses, and water flowing down them attains such high velocity that it toboggans far up the outside walls of meander curves, completely abandoning the floor of the channel. It is sobering to toss a large stone into such a channel and watch it being whisked away by the torrent. There is nothing on the

45

Figure 2.18. Melting of deteriorating glaciers produces a lot of water, which usually finds its way to the bottom of the glacier and ultimately emerges from a subglacial tunnel at the glacier's edge. We here see such a stream, muddy and carrying many icebergs, emerging from beneath the snout of stagnant Steele Glacier in the St. Elias Range. Ice (I) is mantled by a thin layer (arrows) of ablation debris (A), and the black area under (T) is the mouth of the subglacial tunnel. Such streams play an important role in disposing of both water and debris derived from the glacier. (August 1941.)

smooth, slippery ice surface of the channel to impede the stone's movement – or that of a human body, if anyone happened to slip in.

Many such streams disappear down a large, vertical, cylindrical hole in the ice, a **moulin**. It's less than comforting to stand on the edge of such a moulin listening to the roar of the disappearing water as it abandons its **supraglacial** position to follow an **englacial** passage somewhere within the glacier. Eventually the water of the moulin-ingested stream may work its way to the base of the glacier where it becomes **subglacial**. Subglacial channels are largest and most numerous near the outer margin or end of a glacier, where the ice is thinnest. In such settings, one or more torrential streams commonly emerge from large, subway-like ice tunnels (Fig. 2.18).

Most warm glaciers of any size have a complex internal plumbing system consisting of such englacial and subglacial passages. Some englacial systems are tightly confined and operate under pressure, like domestic water pipes. Where they bring water back to the glacier's surface, it may emerge with enough force to form a bubbling spring (Fig. 2.19), or in some instances a geyserlike eruption. Glacier plumbing systems are sub-

Figure 2.19. Glaciers have extensive and complex internal plumbing systems. This small "bubbling" spring of muddy water on the surface of the stagnant reach of Steele Glacier, Yukon Territory, taps a source of meltwater confined under pressure in a closed englacial or subglacial passage, much like a pipe in a domestic water system. The fountainlike water dome is about 35 cm high.

ject to frequent change because of ice movement, and abandoned surface channels and moulins no longer leading to englacial passages are common. Englacial and subglacial channels are kept open by the frictional heat of the water flowing through them. Only the largest are able to maintain themselves against the flow of ice, and once abandoned they are quickly closed. Abandoned moulins on the ice surface usually become sealed at the bottom by ice flow and fill with meltwater.

ORGANISMS AND GLACIERS

Several species of small animals and plants actually inhabit glaciers. Larger animals such as human beings, bears, penguins, seals, gulls, and other types of birds generally only travel across or set up temporary house-

47

keeping on ice. Antarctica may eventually prove to be an exception for humans.

Ice worms, or glacier worms, have long been regarded as a joke in Alaska, foisted upon unsuspecting *cheechakos* (tenderfeet). At one time, Alaskan stores offered a postcard showing a pile of cooked spaghetti lying on snow, captioned "ice worms." **Ice worms**, however, actually exist and live largely within firn, or in pools of water on glaciers, not only in Alaska but in other areas as well. They are small (0.25 mm in diameter and 3 cm long) and black, and crawl through openings between grains in firn or crystals in partly thawed ice. What they live on is not obvious, but they may feed on algae whose airborne spores are caught in snow. Warm firn may also harbor small, black spring-tail bugs, called **glacier fleas** because of their irregular hopping habit. These bugs, too, may feed on algae or perhaps on deceased worms.

Snow and firn on glaciers support colonies of algae that make surficial, dirty greenish spots during summer. These spots turn bright red as the algae pass into a resting stage in late summer or autumn; this is the cause of the red snow well known to mountaineers.

Birds in transit across glaciers feed on both worms and bugs. Birds of many varieties travel extensively over large areas of snow and ice, and a number perish in the process. A person wearing something red can expect to be investigated by hummingbirds in the middle of a vast wasteland of ice and snow many kilometers from any sustenance.

WHERE ARE THE WORLD'S GLACIERS?

Even though ice is the most widespread of all rocks on the earth's surface, it covers only about 3.1 percent of the globe, 10.8 percent of the land. During the last ice age, glaciers covered about 32 percent of the land. Sea ice currently covers 7.3 percent of the ocean. The total volume of existing glacier ice is approximately 33 million cubic kilometers (km^3), but that constitutes less than 2 percent of the planet's water. If all existing land ice were melted, it would raise sea level about 70 meters worldwide. The formation and melting of ice age glaciers caused worldwide sea-level changes of 100 to 140 meters.

Of existing ice the antarctic sheet constitutes 86 percent of the world's total by area (13.9 million km^2), about 90 percent by volume (about 25 to 30 million km^3). Another 11.6 percent, roughly 1.7 million km^2, covers most of Greenland. Not many people are likely to see those ice masses or the 2.7 percent of Earth's ice that lies on islands in the Arctic

Ocean, such as Baffin, Spitsbergen, Novaya Zemlya, and the Queen Elizabeth group. Some of us have seen or will see at least part of the 12,000 km² of ice capping Iceland.

The best opportunity for most people to become acquainted with living glaciers, principally streams of ice, lies in the Alps (2,900 km²), Scandinavia (3,100 km²), Alaska (74,700 km²), parts of Canada (25,000 km²), the Caucasus (1,800 km²), and the Himalayas (33,000 km²). The mountains of South America sport a creditable 36,000 km² of ice, New Zealand about 1,000 km², and mountain ranges of south-central Asia, including the Karakoram, K'un Lun Shan, Hindu Kush, and Pamir, about 109,000 km².

Central Siberia, similar to interior Alaska, however, has surprisingly little glacier ice; it is too dry. Africa harbors only 12 km², about on a par with Mexico, mostly on high volcanic peaks, which also account for New Guinea's 15 km². Scattered glaciers in mountain ranges bordering dry areas in central Asia, such as the Pamir, Tyan' Shan', and others, are much appreciated as reservoirs of precious water.

Although the small ice caps and ice streams in mountainous areas of the northern hemisphere constitute only about 4 percent of the world's ice, they contain about 10 million gallons of fresh water for each citizen of the world. These small glaciers respond quickly to changes in the climatic environment, and their expansion or contraction can cause small but damaging changes in sea level within decades.

CHARACTER AND DYNAMICS OF CONTINENTAL ICE SHEETS

We have focused primarily on valley glaciers because they are the type most likely to be met by readers of this volume. Few of us will see the Antarctic or Greenland ice sheets, except possibly from a passing aircraft, but many of us live in areas formerly inundated by the great North American and Eurasian **continental ice sheets**.

Many of the principles discussed in connection with valley glaciers apply to continental ice sheets, but there are significant differences, especially in dynamics and behavior. A continental ice sheet, to a significant degree, creates its own weather, which naturally is favorable for glaciers. Indeed, a huge ice sheet can have worldwide climatic effects.

Ice sheets become thick enough to submerge the underlying terrain and establish their own independent form rather than being totally controlled by topography, as in the case of a valley glacier. High relief in

49

the terrain underlying an ice sheet is reflected in subdued ice surface relief and currents of faster-flowing ice. Continental ice sheets tend to develop slowly, but once the tide turns they can recede relatively rapidly, sometimes almost catastrophically. Terrain plays an important role in both development and disintegration.

Much is currently being learned about the nature and behavior of ice sheets from glaciological studies in Greenland and particularly in Antarctica. Because ice covers essentially the entire Antarctic Continent, the sheet is tidal around most of its margin and calving is the major form of wastage, especially from the many large, partly floating ice shelves that rim the continent. Huge tabular bergs break from these shelves to drift north and melt in warmer seas. The northern hemisphere continental ice sheets had limited tidal margins, much of the outer edges of the sheets being land based. A further difference is the polar location of Antarctica, which must be considered in extrapolating comparisons to the Pleistocene sheets.

The ice sheet of East Antarctica is a better analogy to the Pleistocene sheets because it occupies a land surface of relatively modest relief, like the interior of North America and Europe. The West Antarctic sheet appears to have submerged an island archipelago of high, rugged relief. Only peripheral parts of the North American and Eurasian sheets submerged such rugged terrains in near-coastal locations.

The behavior of the West Antarctica sheet – the predominance of calving, abundance of fringing ice shelves, and internal currents of fast-flowing ice – is probably different from that of many of the Pleistocene ice sheets. In terms of what most readers will experience, the dynamics and behavior of valley glaciers provide a reasonable background for understanding the behavior of the margins of the Pleistocene continental ice sheets over the areas in which they may live or visit. This is especially true of the landscape features of erosion and deposition treated in Chapters 6 and 8.

RECAPITULATION

Glaciers mostly come in the form of sheets and streams. The sheets can be large, and that of Antarctica contains 90 percent by volume of the world's ice. Glaciers at the melting point throughout, therefore in thermal equilibrium with liquid water, are called warm, and those below the freezing point throughout are cold. Warm glaciers are vastly more dexterous.

Most valley glaciers are compound, being made up of multiple ice streams from tributary valleys separated by septa of dirty ice that underlie medial moraines on the glacier's surface. In some glaciers, such moraines are complexly deformed.

Through flow, glaciers develop small-scale lenticular banding (foliation) like other metamorphic rocks. Foliation displays different strengths and configurations in various parts of a glacier. Icefalls create prominent structures in ice streams, one of the most striking of which is the repetitious curved banding of ogives. In the next chapter we explore the intriguing and sometimes unexpected rates, mechanisms, and results of glacier flow.

3

Glacier movement

One of the more interesting aspects of glaciers is their movement, a behavior we have tacitly assumed in preceding discussions. Most glaciers move by internal solid flow and by slipping over their bed. Ice masses flow across flat terrain as well as downslope, but only after they have attained a size, thickness, and configuration adequate to generate enough differential **stress** to cause the ice to flow as a solid or to slip over its base. Snow and ice accumulating on a mountainside do not flow until the thickness exceeds about 60 meters.

RATES OF MOVEMENT

In *A Tramp Abroad* Mark Twain made the point that most glaciers move very slowly – 1 meter a day is fast for a glacier – by describing a fictitious camp that he and friends had supposedly pitched on an alpine glacier in expectation of a free and joyous ride down the valley, inspecting the landscape as it whizzed by. They were bitterly disappointed when, day after day, the view from camp remained unchanged.

Most glaciers normally move too slowly for direct visual detection, and different parts move at different rates. Mark Twain's group could have learned this by laying boulders in a straight line across their glacier from wall to wall. Curious people did this in the Alps, and probably elsewhere, at least as early as the 1800s. In a year or two the line would no longer be straight. It would be displaced and bent downstream, demonstrating that the center of the glacier, where the ice is thickest, moves more rapidly than the margins (Fig. 3.1). Surface velocity at a point on a glacier represents the cumulative movement of all the ice below that point, hence it is greatest where ice is thickest, other things being equal. If Twain's group had exercised the foresight of establishing

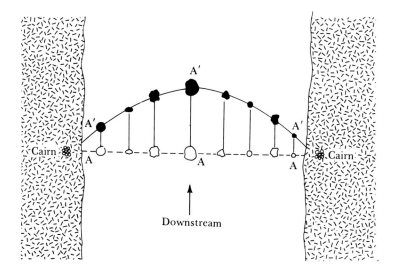

Figure 3.1. Simple experiments performed in the field can yield useful data, as shown by this diagram of a transverse velocity profile (A' A' A') defined by displacement of stones set along an initially straight line (A A A) on the surface of a valley ice stream, as viewed from above. This sort of arrangement was probably one of the earliest experimental observations made of glacier movement, and it revealed that the ice stream flowed more rapidly in its center, where thickest, than along the margins.

cairns on the valley walls, marking the ends of the boulder line, they would also have learned how far the entire glacier had moved.

This experiment has been repeated countless times with greater accuracy and precision using modern surveying techniques. The results are displayed in Figure 3.2 and explained in its caption. As one might expect, variations and complexities are revealed by the more sophisticated surveys, but basic relationships established by the displacment of a line of boulders are shown to be sound.

INTERNAL FLOW AND BASAL SLIP

Suppose we went to the middle of our glacier, drilled or melted a small-diameter hole vertically through to its floor, installed a bendable pipe in the hole, and then surveyed the inclination and position of that pipe for several years. We would find that it not only moved down the valley but was bent into a curve by greater movement at the top than the bottom (Fig. 3.3). We would see, further, that the total down-valley movement of the pipe's top was not fully accounted for by the bending: The glacier had also slipped over its bed (Fig. 3.3).

The ratio of **basal slip** to internal flow varies greatly among glaciers:

53

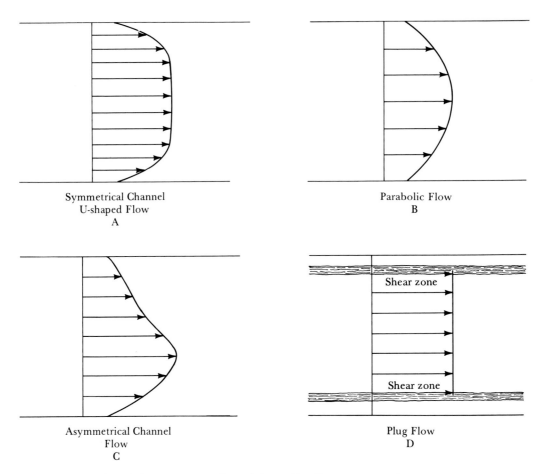

Figure 3.2. Transverse surface-velocity profiles of ice streams recorded by accurate surveys reveal more than the simple line of stones of Figure 3.1. Most glaciers flowing at reasonable speeds in symmetrical U-shaped channels display a broad, open, U-shaped profile like (A), rather than a parabolic profile (B). If the deepest part of the subglacial channel is not in the center, the profile may be distorted (C). Glaciers flowing at abnormally high velocities may experience plug- or block-flow (D), in which the central part of the ice stream moves uniformly as a block between two marginal shear zones. In parabolic flow, the ice is moving slowly and behaving more like a viscous fluid; in U-shaped flow, the velocity is higher and the ice behaves as a quasi-plastic solid; and in plug-flow, at extremely high velocities, it behaves as a perfectly plastic substance (see Fig. 3.10).

Thin glaciers on steep slopes may move largely by basal slip, and thick glaciers on gentle slopes usually exhibit a greater proportion of internal flow. At one time, the ratio for most glaciers was thought to be about 50:50, but modern studies suggest that basal slip usually contributes more than half the movement.

Warm glaciers move more rapidly and display greater variations in velocity than cold glaciers, mostly because their basal slip occurs more easily. Even though cold glaciers are frozen to the bed, they can undergo

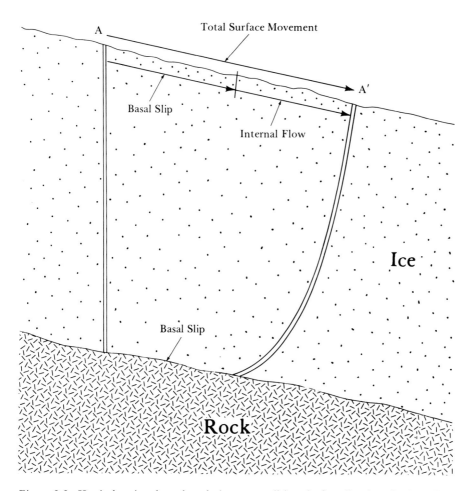

Figure 3.3. Vertical section through a glacier cut parallel to the flow direction, showing what a pipe, initially extending vertically from the surface to the base, would look like after a year or two. The total movement, A A', measured on the surface is the summation of flow within the ice and slip upon the base. Experiments of this type have been performed on a number of valley ice streams.

some localized basal slip under special circumstances. Warm ice also deforms more easily than cold ice, further contributing to the differences in behavior.

Mechanics of basal slip

The bedrock floor of a glacier is usually not a mirror-smooth surface; rather, it is typically endowed with little projecting knobs and ridges, with intervening depressions (Fig. 3.4) that severely retard the rate of basal slip. Even though the protrusions may be rounded and smoothed, their size (from centimeters to several meters) makes the surface rough

Figure 3.4. Glaciologists now recognize that the nature of the bed under moving ice strongly influences the behavior of a glacier. For that reason, fresh exposures of the bed, as shown here just beyond the receding snout of Blue Glacier (background), are of great scientific interest. This glacially plucked, smoothed, striated, and polished rock surface is probably rougher, harder, and more impermeable than many glacier beds. The numerous small protuberances must have significantly retarded, but clearly did not prevent, basal slip. The erratic boulder (right center) is about 50 cm high. (Photo by Barclay Kamb, summer 1972.)

to the glacier. The basal ice makes its way over and around these irregularities, either by melting and refreezing or by solid flow. Both processes play a role, melt-and-freeze mostly at the smaller obstructions and solid flow mostly around the larger ones. Both act slowly, and hence basal slip is retarded. This is just as well, because our warm glacier with a film of water at its base would slip so easily over a perfectly smooth bed that even Mark Twain and his party would have been pleased with the rapidly passing scenery. The glacier would dispose of its supply of ice so quickly that it could never grow to any significant size.

Physicists long ago coined the term **regelation** for the melting of ice under increased pressure and the refreezing of meltwater upon reduction of pressure. This happens as a glacier overrides small obstructions on its bed: The ice melts on the upstream side and refreezes on the downstream side. The layer of ice (usually a few centimeters thick) affected by this process is the **regelation layer**. It is distinguished by thin lamination, small crystal size, and abundant rock debris. Regelation layers are formed along the base of warm glaciers. A cold glacier has to be almost warm to develop even a weak regelation layer. At temperatures colder than $-22°$ C, no amount of pressure causes ice to melt; it simply changes to a denser, high-pressure form of ice.

Direct visual inspection of conditions at the base of a glacier is possible in places under the thin ice of an icefall or along a glacier's margin. Modern techniques such as downhole photography to determine bed constitution and roughness, as well as measurements of variations in water pressure at the bottom of boreholes, have also been employed under thicker ice. Such data show beyond doubt that conditions at the ice base play a major role in glacier behavior. Our understanding of erratic glaciers has advanced with the realization that warm ice is not everywhere in firm contact with its bed but rather is locally held off by pockets of water in small cavities.

DIRECTION AND AMOUNT OF MOVEMENT

Let us return to a consideration of what happens annually within the accumulation and wastage areas of a valley glacier. Every year a wedge-shaped layer of snow, thickest at the head and thinning to a feather edge at the snowline, is added to the surface of the accumulation area. A differently shaped but similar wedge of ice, thickest at the snout and thinnest at the equilibrium line, is removed by melting from the wastage area (Fig. 3.5). Assume that the glacier is in a steady state, neither shrinking nor expanding; then the volumes of water represented by these two wedges must be the same.

Now the glacier has a problem. As a stable, steady-state ice stream it must preserve its geometrical form by transferring ice in such a way as to maintain a stable longitudinal profile. It cannot keep getting thicker at the head and thinner at the snout. Having the greatest velocities of flow at the head and snout, where the largest amounts of material are being added and lost, seems like a possible solution. Measurements of velocity show that this is simply not so. If the glacier has a channel of reasonably uniform shape and size and if the down-valley slope is uniform, the greatest velocity must be at the firn line. This is so because ice

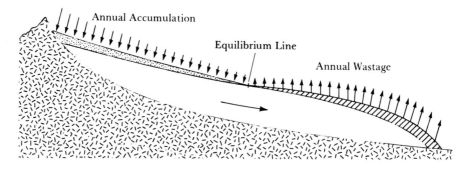

Figure 3.5. The wedge shape of annual layers of accumulated and wasted material on the surface of a valley glacier, shown here in a diagrammatic, two-dimensional longitudinal section from the head to the snout, creates problems. The glacier cannot simply go on getting thicker at the head and thinner at the snout. It adjusts by an internal flow arrangement depicted in Figure 3.6. If the glacier is in an equilibrium state, the volumes of water represented by the accumulation and wastage wedges, in three dimensions, must be the same.

thickness and discharge increase from the head to the firn line and decrease from there to the snout. Velocity is greatest where thickness and discharge are at their peak, icefalls excepted.

The glacier solves the problem of adjusting to the accumulation and wastage wedges by changing the inclination of the direction of flow. At the firn line, the direction is parallel to the surface of the glacier. In the accumulation area, it angles downward with respect to the surface, and in the wastage area upward. The degree of downward inclination *decreases* from the head to the firn line and the upward inclination *increases* from firn line to snout. The downward or upward movement is measured with respect to the sloping ice surface. Upward movement of ice with respect to the horizontal can occur in glaciers, but it is not required to make these adjustments.

Suppose we cut our valley glacier in half along a vertical plane extending from head to snout and remove one half to expose the interior. Then the **vectors** of ice movement (arrows whose lengths measure velocity and whose orientations indicate direction) would look approximately like those in Figure 3.6. Glaciers flow in a **laminar** manner, meaning that lines or surfaces of flow are conformable and do not intersect or cross. Ice flow, unlike the flow of water in rivers, is free from turbulence or mixing of adjacent layers. Adjacent vectors of flow in any single section through a glacier differ in length (velocity) but not drastically in orientation (direction). The tendency toward **turbulent flow** in any fluid increases with velocity but decreases with viscosity. Someone once calculated that because of the high **viscosity** of ice, glaciers would have to flow with velocities approaching the speed of light before turbulence

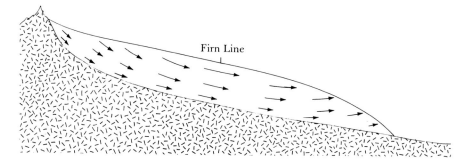

Figure 3.6. *The arrows in this longitudinal section through a valley glacier are vectors; their orientation indicates the direction of movement, and their length represents the relative amount of movement on an arbitrary scale. They are longest and parallel to the glacier's surface at the firn limit and decrease in length toward both the snout and the head of the ice stream, where the lowest velocities prevail. In any selected cross section, they are also shorter toward the base of the glacier, because arrows closer to the surface represent the cumulative movement of all the underlying ice. Note that the arrows are directed downward from the glacier's surface above the firn limit and upward below it. These directions of movement permit the glacier to adjust to the wedge shape of the annual layers of accumulation and wastage depicted in Figure 3.5.*

would occur. This does not mean that parallel laminae of flowing ice are smooth, flat planes. They can be curved, but they don't mix.

The flow lines shown in Figure 3.6 give rise to some interesting interpretations. For instance, ice composing a glacier's surface just below the firn line should have been made from snow that accumulated just above the firn line, so it should be young, immature glacier ice. This is confirmed by its crystal size, high percentage of air bubbles, and lack of strong secondary structures created by flow. In comparison, the larger crystals, paucity of air bubbles, and strongly developed secondary structures of ice at the snout suggest it is old. If the arrows of Figure 3.6 are extended into continuous flow lines, they show that ice at the snout comes from the head of the glacier and was near the bottom as it passed the firn line. It has had a long, hard trip. (See the boxed explanation of oxygen-isotope ratio measurements on pages 66 to 67 to determine the source of ice composing a glacier's snout.)

EXTENDING AND COMPRESSING FLOW IN GLACIERS

Because velocities in the accumulation area increase progressively from the head to the firn line, down-valley ice is consistently pulling away from up-valley ice throughout the accumulation area. This is the condition of **extending flow** (Fig. 3.7). Theoretical models of its character-

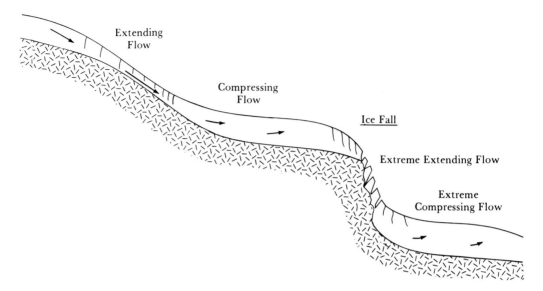

Figure 3.7. Where velocity accelerates, a glacier is extended and thinned; where velocity de-celerates, a glacier is compressed and thickened. The ice itself suffers neither elastic expansion nor compression. This diagrammatic longitudinal profile down the center line of an ice stream with reaches of different steepness shows how extending flow rules in thin, fast-flowing ice over steep descents and compressing flow occurs where the glacier slows up within reaches of gentler gradient.

istics and behavior correspond with phenomena observed on glaciers, such as crevasses and the downward orientation of flow vectors. Below the firn line, velocities progressively decrease, and up-valley ice is continually pushing against down-valley ice. This is the condition of **compressing flow** (Fig. 3.7). It creates flow vectors directed surfaceward that produce a thickening of the glacier to counterbalance increased melting. This interesting mechanism meets the budget needs of valley glaciers.

Natural channels are seldom uniform in size, shape, and gradient. Almost always, reaches of gentler and steeper slope (**topographic gradient**) alternate along a valley. Where the slope steepens, glacier velocity increases and extending flow prevails; where the slope gentles, velocity decreases and compressing flow occurs (Fig. 3.7). Some glacier channels have abrupt steps over which the ice pours with high velocity, creating a veritable fall of ice (Fig. 2.9). These are reaches of extreme extending flow in which the ice is greatly thinned and completely crevassed, creating sharp-crested angular blocks known as **seracs**. The flow velocity in an icefall can exceed ten times that of the glacier elsewhere along its course.

At the base of the fall, conditions are reversed; velocity decreases rapidly, the glacier thickens greatly, and intense compressing flow reigns. Like streams of water, glaciers are thinner (shallower) in reaches of high velocity and thicker (deeper) in reaches of slower velocity, the thinning and thickening being accomplished by the mechanisms of extending and compressing flow.

Planimetric directions of ice flow, as plotted on a map, are controlled primarily by the slope of the glacier's surface. Inclination of the bed usually plays a major role in determining that slope, but the two are not always fully concordant. In an ice sheet the flow is principally radially outward from the center or from broad domes on its surface. Movement near the base, however, is strongly influenced by the local configuration of the subglacial topography. A buried preglacial valley will channelize the flow of basal ice. Along a transverse profile of a valley glacier below the firn line the ice surface slopes downward toward the margins in opposition to the upward slope of the bed. The surface slope is produced by greater melting toward the margins caused by higher air temperatures and by radiated and reflected heat from the valley walls. This outward slope generates a small component of marginward flow, which is manifest in the slight splaying of arrows depicting planimetric flow directions in a typical valley glacier (Fig. 3.8).

SOLID FLOW OF ICE

Given adequate time and stress, many substances normally regarded as solids can be made to deform permanently without fracturing. Gravestones hundreds of years old have bent and drooped. Mention was made in previous pages of the ability of ice to flow as a coherent, integrated solid. This is demonstrated in the laboratory where solid ice under pressure flows through a pipe or is extruded from an orifice. The broad, graceful curves of medial moraines in the Kaskawulsh Glacier (Fig. 3.9) leave little room for doubt that glaciers flow readily. Ice flows most easily in a near–0° C environment, but even at tens of degrees below freezing it can be made to flow, albeit more reluctantly.

Air and water are perfect **Newtonian fluids**, which obey a simple **flow law** represented by the inclined straight line, *A*, in Figure 3.10. This line shows that increased **shear stress** produces increased rates of response at a fixed ratio governed by the constant viscosity of the substance. The line for air is less steep than the line for water, but both are straight and both start at the origin, or zero point, on the shear stress–response diagram (Fig. 3.10). In these substances, the very smallest stress produces some flow. **Perfectly plastic substances** obey a different law represented by the straight horizontal line, *B*, in Figure 3.10. They show no nonelastic yielding whatever under increasing stress up to a point, after which they yield so readily that shear stress cannot be increased. The stress at which

61

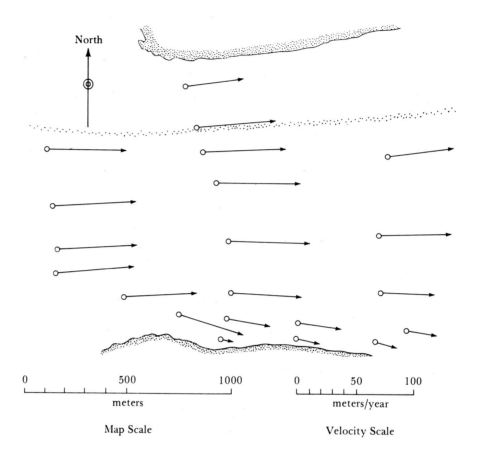

Figure 3.8. These arrows represent the movement of stakes set into the surface of Saskatchewan Glacier, in the Canadian Rockies, as determined by accurate surveys. They show that the down-valley flow is slightly dispersive toward the glacier's margins, especially on the south side. This happens because of the outward slope of the ice surface toward the margins owing to accelerated melting there. (After M. F. Meier, 1960, Mode of Flow of Saskatchewan Glacier, Alberta, Canada. U.S. Geological Survey Professional Paper 351, 70 pp.)

they first start to respond is called the yield point (K in Fig. 3.10). A whole group of solids, including ice and many other rocks, yield according to a **viscoplastic** or **quasi-plastic** flow law represented by line C in Figure 3.10. They differ from Newtonian fluids in showing no flow over short time intervals under low shear stress, but once they start to yield, at higher stresses, they initially behave something like a Newtonian fluid in that an increasing stress produces higher flow rates, but not at a fixed ratio. Eventually, at high shear stresses the rate of yielding becomes so great that curve C begins to flatten out like line B of a perfectly plastic substance.

Let us accept the principle that ice partakes of solid flow and address the means by which it does so. Figure 3.11 depicts some of the ways by

Figure 3.9. This view up Kaskawulsh Glacier in the Yukon's St. Elias Mountains shows how flowing ice can adapt gracefully to a winding course, provided its channel is wide and the curves not too sharp. Remember that the dark medial moraines are simply residual accumulations of rock debris on the glacier's surface marking the traces of underlying septa of dirty ice (see Fig. 2.5). Note how the tributary entering from the left is compressed into a narrow band as it is squeezed between the trunk glacier and a large, vigorous (crevassed) branch coming in from the lower left. (U.S. Geological Survey photo by Austin Post, August 26, 1960.)

which a block of ice might yield to stress and yet maintain its solidity and integrity.

If the crystals of ice composing a glacier were only loosely bound, like grains of sand on the beach, the glacier could flow simply by allowing the grains to move past each other (Fig. 3.11A). Glaciers are not like this, however. It is beyond doubt that ice crystals in glaciers are tightly and

63

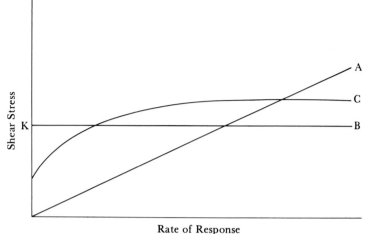

Figure 3.10. This shear stress–rate of response diagram shows that a Newtonian viscous fluid (A), such as water, responds in straight line fashion to increasing shear stress: The greater the stress, the greater the rate of response at a fixed ratio. Leaving aside elasticity, a perfectly plastic substance (B) does not yield to shear stress until a certain threshold value (K) is reached; then it yields so readily that shear stress on it cannot increase. A quasi-plastic (or viscoplastic) substance (C), such as ice, has a lower yield point than a perfect plastic, after which it yields somewhat like a viscous substance but at a rate increasing more rapidly than shear stress, so the graph line is curved. Eventually, when shear stress is high, ice behaves almost plastically.

complexly intergrown, like a three-dimensional jigsaw puzzle. We know that temperate ice melts under pressure. Perhaps stress exerted on an aggregate of complexly intergrown, irregular ice crystals causes them to melt at points of concentrated pressure, with the water moving to places of lower pressure and refreezing (Fig. 3.11B). This probably happens, but unless orientation of the stress is constantly changing, it is a self-limiting process and will not work in cold ice, where the pressures required are too high.

We saw earlier that glacier flow is laminar, so we might expect that thin layers of ice slipping past each other like playing cards in a deck could result in flow (Fig. 3.11C). This does occur to some degree in thin, brittle bodies of ice and in near-surface crusts of glaciers, allowing them to deform. The development of shear-type foliation probably involves this sort of differential displacement. Some ice that has flowed shows no signs, however, of laminated structure.

The currently preferred explanation for most solid flow in glaciers involves the behavior of individual ice crystals. It is known from laboratory experiments that single ice crystals can be deformed by slippage on a host of parallel **glide planes** of preferred crystallographic orientation,

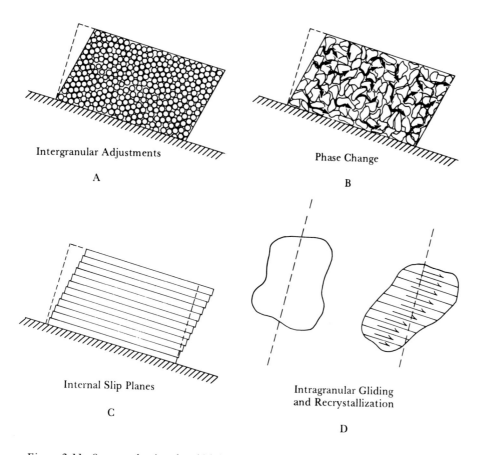

Figure 3.11. Some mechanisms by which ice may undergo solid-state flow. (A) In ice composed of loose individual granules, shifting of the grains can produce flowlike movement. (B) Ice composed of interlocked crystals could experience similar shifting and movement owing to local melting, migration of the fluid (black arrows), and refreezing (regelation). (C) Small displacements along a series of parallel slip planes within an ice mass can produce significant cumulative movement. (D) Displacements on parallel glide planes within an ice crystal are known to occur under stress, producing movement and deformation of the crystal, which must ultimately regain an undistorted shape by recrystallization because ice crystals in glaciers are not sheetlike.

somewhat like a stack of playing cards. This is the process of **basal gliding** (Fig. 3.11D). If long continued, it would distort an equidimensional crystal into a sheet, but crystals in glaciers are not sheetlike. In cooperation with adjacent crystals they seem to be able to reestablish an equidimensional shape by recrystallization, after deformation through internal gliding. In the process of reorganization the crystals acquire a preferred orientation favorable for gliding, creating what crystallographers call a **crystal fabric**, and such **fabrics** have repeatedly been found in glaciers.

Solid flow in glacier ice seemingly involves continuous reconstitution

One way to determine whether ice at the glacier's snout actually formed from snow accumulated near its head is by measuring its oxygen-isotope ratio in a **mass spectrometer**. High school chemistry taught us that normal oxygen has an atomic weight of sixteen (written ^{16}O). Other types of oxygen, called **isotopes**, have an additional **neutron** or two in their nucleus, which increases the atomic weight to seventeen or eighteen. Most natural water contains about 0.2 percent ^{18}O, and the remaining oxygen is almost entirely ^{16}O. All of us have some ^{18}O and lots of ^{16}O in our bodies. (Neither is radioactive, so don't worry; the ^{18}O simply makes us trivially heavier.)

When water vapor forms by evaporation from any water body, it leaves behind some of the heavier ^{18}O atoms. Thus, the vapor is impoverished in ^{18}O compared to the water source. As some of the vapor subsequently condenses to rain or snow, the remaining vapor is further depleted in ^{18}O, as shown in the cartoon below. The mean **oxygen-isotope ratio** ($^{18}O:^{16}O$) of ocean water is taken as a standard, or zero point,

on the ratio scale. Departures from that value can be positive or negative and are expressed as a deviation (δ) in parts per thousand. The δ value of initially evaporated vapor is -7. Much natural rain and snow have δ values between -15 and -30. It has been shown that higher elevation, higher latitudes, and colder conditions all favor more negative values. The lowest yet measured, -62, is in ice from a deep borehole near the middle of the high, cold East Antarctic Ice sheet.

On a valley glacier, ice formed from snow accumulated near the head, which is higher and colder than the lower reaches, should have a more negative $^{18}O:^{16}O$ ratio than snow formed under warmer conditions closer to the firn line. This relationship can be disrupted by local environmental aberrations such as wind-drifting of snow, but the principles hold. Actual analyses have shown that ice at a glacier's snout does have a more negative $^{18}O:^{16}O$ ratio than ice just below the firn line as the flow lines (Fig. 3.6) suggest it should.

and reorganization of individual crystals, and in the process, slow as it is, the ice mass deforms, adjusts, and flows, maintaining its solid state. Most mineral crystals, including ice, have internal imperfections known as **dislocations**, which, under stress, move around within the **crystal lattice** and play a part in causing it to yield. This complex process, termed **dislocation creep**, is not yet fully and satisfactorily understood.

RECAPITULATION

Most parts of most glaciers move slowly, a fraction of a meter per day. Velocity measurements on the surface represent a summation of what is

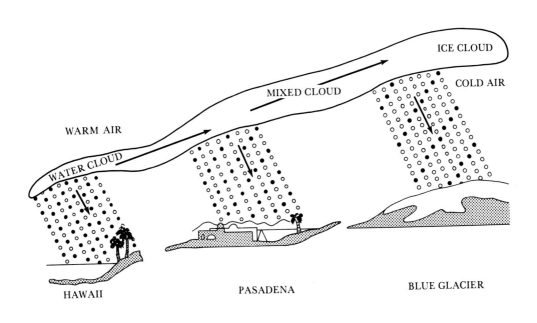

This cartoon shows that rain falling on warm, humid Hawaii contains more of the ^{18}O isotope (black dots) than rain falling on cooler, drier Pasadena, California, which in turn is richer in ^{18}O than snow falling on the colder Blue Glacier of Mount Olympus, Washington. The ^{18}O isotope is slightly heavier than normal oxygen, ^{16}O, so precipitation relatively impoverished in ^{18}O, such as the snow of Blue Glacier, is said to be lighter than the rain at either Pasadena or Hawaii.

happening by solid flow within the glacier and by slippage over its bed. Even though basal slip is retarded by unevennesses of the bed, it contributes the larger fraction of the total movement of most glaciers. Warm glaciers move more rapidly and experience greater velocity changes than do cold glaciers.

An ice stream flowing in a uniform channel has its greatest thickness, highest velocity, and largest discharge at the firn line. Ice flow is laminar, not turbulent, and is directed downward from the surface in the accumulation area and upward in the wastage area. As a result, ice at a glacier's snout comes from the head of the glacier, not from some intermediate position.

Where velocity accelerates, extending flow occurs, and a glacier thins. Where velocity decelerates, compressing flow occurs, and a glacier thickens. An icefall is a reach of extreme extending flow, and the plunge pool at its base is a realm of extreme compressing flow.

Most glaciers, especially warm ones, do not have steady velocity regimes. They undergo subtle to major velocity changes of diurnal, seasonal, or secular frequency. Some secular changes can be catastrophic, with velocities temporarily attaining 100 meters per day. Such surges and other intriguing glacier behaviors are treated in the following chapter.

4

Unusual glacier behavior

Imagine that we have set up a highly precise and accurate system for measuring variations in flow velocity along a series of transverse profiles across our glacier, employing the most modern surveying techniques. Assume further that it will operate continuously throughout the year repeating observations every two or three hours. This is a demanding experiment, but it should produce interesting results.

SOLITARY WAVES IN GLACIERS

One of the first discoveries will be that the flow velocity usually peaks in spring. If our measurements are truly precise, we might also record a diurnal velocity fluctuation, the maximum occurring in late afternoon or early evening. We might also discern a difference between warm and cool days or periods. Other more irregular secular velocity variations of longer duration and greater magnitude will also be recorded.

If we are lucky, we might stumble onto a condition in which velocity increases steadily from one day to the next for a number of days until it attains a value several times greater than the usual fraction of a meter per day. Subsequently, it might return to normal or stay at the higher value. If we have been concurrently observing elevations of the ice surface along our transverse lines, we will see that the surface has risen and then settled back as the velocity returned to normal, or it may stabilize at the new level, if the velocity remains high. When data from all transverse profiles are assembled and analyzed, we will realize that a wave or bulge has passed through the ice. This phenomenon has been characterized by some glaciologists as a **kinematic wave**.

Several varieties of kinematic waves exist, but the type postulated in glaciers is a solitary wave of increased but constant discharge, involving greater velocity and ice thickness, that moves down-valley through an ice stream or outward within an ice sheet. We are familiar with compressing and extending flow within glaciers. A glacial kinematic wave might be pictured as a moving zone of compressing flow causing an increase in thickness, followed by a zone of extending flow causing an increase in velocity. If we install a noise detector in our glacier and happen to be listening as the wave passes, we will hear a large increase in the creaking, cracking, and groaning that accompany an increase in velocity.

Not all nonglacial kinematic waves operate in an advancing mode; some are fixed, and some are regressive. A jam-pack of bumper-to-bumper traffic on an expressway is a kinematic wave of discharge that many of us have experienced. It is commonly fixed in position: Cars enter, pass through, and with patience and luck emerge at the far end. If cars are being added at the rear faster than they can be conducted through, the wave builds up regressively until a balance is struck between cars entering and leaving; then the wave becomes stabilized.

Kinematic waves in glaciers can be generated by successive years of increased accumulation, and they can be the means by which the glacier tells its snout to advance in order to bring the budget back into balance. An increase in discharge amounting to only 5 percent at the firn limit becomes progressively larger, in a relative sense, within the thinner, slower-moving ice approaching the snout. Whereas the increase in thickness at the firn limit may be a meter or less, it can be many tens of meters near the snout. The effect is like cracking a whip. A small flick of the wrist at one end produces a major flap at the other. The glacier snout responds dramatically in terms of increased thickness, flow velocity, and rate of advance.

The concept of waves or bulges passing through glaciers helps explain many aspects of their observed behavior and makes them especially interesting subjects of investigation. Only a few waves have actually been tracked as they moved through ice streams. Signs of their passage can usually be recognized during or after the fact, but more actual measurements of wave character and behavior are needed.

SURGING GLACIERS

In mid-December 1936 occupants of the Rapids Roadhouse, a wayside inn on the Richardson Highway crossing the high Alaska Range, were startled and alarmed upon looking westward across the Delta River Valley. They saw that the normally smooth, gently inclined front of the

3-km-wide Black Rapids Glacier had been transformed almost overnight into an abrupt, steep, strongly crevassed ice face more than 100 meters high. Within a few days it became clear that this formerly receding glacier was advancing rapidly into Delta River Valley. Daily advances as great as 90 meters were subsequently recorded. If continued for long, such an advance threatened to dam the Delta River, block the highway, and demolish the roadhouse. Consequently, this event received much publicity and attention, partly because of the ease of access. Fortunately, the advance ceased after a few months and, at maximum, fell short of any destructive action. Subsequently, the lower part of Black Rapids Glacier stagnated and is now wasting away in place.

This behavior is termed a glacier **surge**, and it has been observed in other places from at least the beginning of the century. Most surges have occurred in remote areas, so few have been documented as well as the Black Rapids advance. One of the most rapid surges on record, 12 km in three months, was by Kutiah Glacier of the Karakoram Range in Asia. A larger surge of 21 km formed the Bråsvellbreen Glacier on the arctic island of Spitsbergen, north of Scandinavia, between 1935 and 1938, but the duration, and therefore its rate, is not known. Surging glaciers have been known to attain velocities one hundred times normal. A fortuitously timed study of southern Alaska coastal glaciers recorded catastrophic advances, presumably surges, of nine ice streams in the Yakutat Bay area between 1905 and 1908.

The snout of the large tidal Hubbard Glacier (Plate I) in this same area has been advancing slowly for a century, and in 1986 it closed off the mouth of Russell Fjord, as predicted by glaciologist Austin Post twenty-five years before. Despite press statements to the contrary, Hubbard Glacier did not surge, although one or more of its tributaries may have experienced small surges. Serious damage to marine life in Russell Fjord and to salmon in Situk River was avoided when the dam of debris and ice failed.

We now realize from observation of chaotically disrupted surfaces, strongly sheared margins, major changes in thickness, highly deformed medial moraines (Fig. 4.1), and extensive reaches of stagnant terminal ice (Fig. 1.10) that many valley glaciers have experienced surges, either in the trunk stream or in tributaries. Well over two hundred surged glaciers have been identified in northwestern North America, although surges have actually been observed in progress in only a few of them. One of the better documented examples occurred in Muldrow Glacier (Fig. 2.6), draining from the east side of Mount Denali (McKinley) some time within a ten-month interval beginning with early May 1956.

Figure 4.1. Medial moraines on glaciers are usually reasonably linear (see frontispiece), but not on Susitna Glacier in the Alaska Range. These have been severely deformed by powerful surges within tributary ice streams of this compound glacier, especially by the branch coming in from the upper left-hand corner, and the tributary from the right near top center. It appears that ice stream A has surged at least twice to create bulbous lobes A_1 and A_2, and ice stream B has possibly surged three times to create areas $B_1(?)$, B_2, and B_3. An alternative possibility is that the trunk glacier has done the surging, shearing off the ends of the protruding tributaries. The way the medial moraine, bounding the right side (looking upstream) of ice stream C, swings across to the lateral margin of the glacier suggests that stream C is inset into stream A. The floor of its valley, at left, may hang above the floor of A. Ice stream C may also have surged at the same time as A_2. (Photo by Bradford Washburn.)

Morainal features in the middle reach moved as much as 6.9 km down-valley, and ice thickness in the lower part increased by 90 meters. The stagnant lowermost reach was reactivated, and the snout advanced. Concurrently, the ice level in high tributaries of the Muldrow system fell abruptly by 40 to 60 meters.

Surge of Variegated Glacier, Alaska

If we wished to be in the forefront of modern glaciological research and involved in a topic of high current interest, we could temporarily abandon our well-behaved ice stream and go in search of a prospective surging glacier. One problem in understanding surging glaciers has been a lack of detailed knowledge concerning them before they surged. Until recently, glaciers had been studied mostly after surging, not before or even during that condition. Should we set out to monitor a surging glacier, we would probably select an ice stream, not too big, situated in an area accessible both summer and winter, with a prior record of periodic or near-periodic surges.

A group of glaciologists from academic institutions in mainland United States, Alaska, and Europe conducted such an endeavor. As their subject they selected Variegated Glacier at the head of Yakutat Bay on Alaska's southern coast, which had a record, starting in 1906, of four surges spaced roughly seventeen to twenty years apart. Considering the time to be ripe for another surge, they initiated exhaustive studies in 1973 extending over a decade to define physical characteristics of the glacier. Such things as ice thickness, temperature, meltwater discharge, channel configuration, structure, and especially variations in velocity regime were determined. Direct surface observation, many boreholes to the bed, and various indirect geophysical sensing techniques were used. Behavior of meltwater within the glacier was given particular attention, especially the hydraulic pressure at the bed. Variegated Glacier became one of the most thoroughly studied warm valley glaciers in all of North America.

The scientists waited patiently, recording and interpreting the glacier's pulse, as it were. Their foresight was rewarded in 1982 and 1983, when Variegated Glacier initiated a number of mini-surges climaxing in a two-phased, pulsing, massive surge that profoundly altered the physical aspects of this ice stream (Fig. 4.2). They literally captured a surging glacier.

The major surge was not a single simple event. Rather, the glacier behaved like an athlete warming up for an event by taking deep breaths, doing calisthenics, and engaging in short sprints to flex muscles and get in trim for the main effort. Many mini-surges occurred before the big surge, and one even took place afterwards, similar to a warming-down exercise. These mini-surges typically lasted a few hours to a few days and generated velocities approaching ten times the usual velocity of less than 1 meter per day. The mini-surges themselves moved through the ice stream at velocities of hundreds of meters per hour, usually building up quickly and tapering off more slowly.

Figure 4.2. In July 1982 a corresponding photo looking up Variegated Glacier, in the Yakutat Bay area of southern Alaska, would have recorded a smooth, subdued ice surface at a level many tens of meters lower than the rough, crevassed surface shown here on July 4, 1983. The disruption and change were caused by a surge culminating on that date and terminating by July 26. This surge was witnessed, literally captured, by a team of glaciologists as the culmination of a ten-year study of Variegated Glacier. At lower right, between the valley wall and the glacier, is a pond (P) filled with floating ice fragments. It developed during surging and drained as the surge died out, suggesting an initial increase and then a decrease in basal water pressure as the normal drainage system of the glacier was reestablished. (Photo by Elise B. Mezger [Weldon], courtesy of Barclay Kamb.)

The major surge occurred in two stages over a period of eighteen months, starting in January 1982. Its second phase, beginning in October, resulted in velocities nearly forty times normal over extensive reaches of the glacier and up to one hundred times locally. The surging climaxed on July 4, 1983, and was essentially over by July 26.

The surge produced as much as 100 meters of thinning in the upper part of the glacier, while the lower part was thickening by a similar amount. An abrupt, steep ice front moved through the lower reach of the glacier leaving an expanded mass of highly crevassed ice in its wake (Fig. 4.2).

Measurements in a borehole to the base of Variegated Glacier showed that fully 95 percent of the movement during surging was by basal slip. Because warm glaciers, in which water and ice can coexist, are the kind known to surge massively and frequently, we might infer that water plays a major role. That thought was certainly in the minds of the Variegated Glacier investigators. They carefully monitored the level of water standing in crevasses and in boreholes, measured the hydraulic pressure at the bed of the glacier, observed changes in amount and places of discharge of water along the glacier's edge, and followed changes in circulation rates within the glacier's plumbing system by means of fluorescent dyes.

Any change that lessens the obstruction of protuberances on a glacier's bed or reduces the amount of contact between ice and bed would accelerate basal slip. It is hypothesized by glaciologists that water gradually accumulates within small cavities under the ice, which are connected with one another and with the normal subglacial plumbing system only by narrow passages that do not transmit water easily. Under such conditions the basal hydraulic pressure could rise to the point at which the glacier is locally raised a centimeter or two off its bed. That would greatly increase basal slip and could initiate and sustain a surge.

This is what the Variegated Glacier investigators think happened. As the surge was waning, abnormally large discharges of water occurred from subglacial streams at the glacier's margin, indicating that the normal plumbing system was draining off the accumulated basal waters, allowing the glacier to settle back onto its bed and terminating the surge. A major enigma remaining is why and under what conditions some glaciers can build up an excess of basal water adequate to generate a surge at widely separated intervals.

One can wonder whether mini-surges, and possibly major surges, too, should be regarded as kinematic waves. Experts are somewhat ambivalent on this matter, partly, one suspects, because the mechanism causing increased flow in kinematic waves is not fully understood. The advancing front of a surge has the characteristics of a kinematic wave, but it may not involve the same mechanisms that produce kinematic waves. The physical laws controlling kinematic waves and surges may be different. We need to remember that the kinematic wave concept is no more than a theoretical model used to explain a glacier's behavior. Glaciologists now know that some glaciers undergo major surges, and perhaps all glaciers have mini-surges.

The history of a glacier after surging warrants attention. Usually, the surface of its upper reaches sinks back to a level lower than before the surge. The lowermost part, which has thickened dramatically and advanced vigorously, stops moving, stagnates, and wastes away in place. As the ice melts, englacial debris accumulates residually on the surface, eventually making a nearly complete mantle, and the only ice visible is in the walls of superglacial ponds and streams. A wasted, **stagnant** glacier is not a pretty sight (Fig. 4.3).

The possibility of surging of tidal glaciers into **fjords**, long, narrow arms of the sea occupying deep glaciated valleys, is of more than just scientific interest. It has long been known that fjords on the coast of Greenland, into which outlet glaciers from the inland ice sheet debouch, periodically become choked with icebergs. These bergs then float into the Atlantic Ocean where they constitute a hazard in shipping lanes. Greenland glaciers discharge on the order of ten thousand icebergs into the Atlantic annually. Although surging could cause a dramatic increase in berg abundance, other influences may actually be dominating. Big outlet glaciers from ice sheets have such a large discharge they don't need to surge to create an oversupply of bergs; Jacobshavn Glacier on the west coast of Greenland is a good example.

In other situations variation in berg abundance is related to the normal cycle of slow advance and catastrophic rapid recession experienced by many tidal glaciers, such as those of Glacier Bay, Alaska. A current example is the large Columbia Glacier in Prince William Sound on Alaska's central south coast. This tidal glacier, a popular tourist sight, is beginning a rapid retreat that is creating many large icebergs. For decades the glacier's terminus has been resting in shallow water on top of a morainal sill at the mouth of its fjord. Now, as the glacier enters a recessional mode and the ice front retreats, water depth is increasing because the fjord bottom falls away under the snout. Eventually, water depth will become great enough nearly to float the glacier, and catastrophic disintegration will occur. Fortunately, the largest bergs go aground on the shallow sill, so they may not drift into the nearby lanes plied by large oil tankers supplied by the Trans-Alaska pipeline. Nonetheless, the pos-

Figure 4.3. Looking down onto the stagnant terminal area of Logan Glacier in the westernmost St. Elias Range, Alaska. This glacier earlier experienced a surge of unknown date, which produced an overextended condition leading to stagnation. It is now recognized that most glaciers with extensive stagnant lower reaches have previously surged. The mantle of rocks, gravel, and sand on the ice is probably less than 1 meter thick. Little bare ice is to be seen. The smooth, oval, light- to dark-gray spots are ponds.

sibility of an oil spill in this biologically rich, sensitive area has understandably put people on edge. The Columbia Glacier has been thoroughly studied and is being carefully monitored.

RECAPITULATION

Glaciers exhibit unusual velocity variations. One common phenomenon is the movement of a wave of increased discharge through a glacier at a speed several times normal. This behavior has been described and analyzed as a kinematic wave. Such waves enable a glacier to adjust to increases in material budget in a much shorter time than that required to transport the additional material to the wastage area.

Many glaciers are known by direct observation, or from circumstantial evidence, to have undergone a major surge advancing the terminus at up to one hundred times the normal flow velocity. Such surges involve a large and rapid increase in basal slip. Field observations suggest that this is owing to an accumulation of water in cavities beneath the glacier, which increases the hydraulic pressure to the point that the contact between the glacier and its bed is reduced. The relationships between kinematic waves and surges are unclear; the mechanisms causing the velocity increase may not be the same.

Now that we are informed on glacial characteristics, features, behavior, and modes of movement, let us see what glaciers can do to the landscape over which they move. This is glaciation, or glacial geology, a matter closer to home for many of us. Chapter 5 introduces the challenging and difficult topic of how a glacier erodes its bed.

5

Glacial erosion

To see how glaciers pick up and carry away rocks, soil, vegetation, or any other material, we must focus on processes and conditions at the bottom of the ice. Preceding chapters have shown that processes at the ice/bed interface play a major role in glacier movement and behavior. Many of the same considerations apply to glacial **erosion**. Essentially all erosion accomplished by any glacier occurs along its bed and the walls and sides of any channel confining it.

Getting to the bottom of a glacier is no easy task. Only a few people have made direct observations of the bed of an active glacier, and then under ice mostly less than 100 meters thick. Access has been gained by descending into crevasses or by penetrating marginal cracks. Small tunnels have been dug along the floors of glaciers by scientists and larger ones beneath glaciers by electrical utility companies. They are interested in gaining access to water that does not freeze in winter within and beneath warm glaciers, which can be used for generating power. Mining companies seeking mineral deposits lying beneath glaciers excavate rock tunnels from which vertical shafts have been opened to the glacier's floor.

Despite these accesses, knowledge of how glaciers erode based on direct observations is distressingly small. Boreholes through ice to the glacier's floor, down-hole photographs, and various geophysical remote sensing techniques are helpful, but at best they give only a bird's-eye view. We have access, fortunately, to some informants that were there while the erosion was going on, participated in the activity, and were excellent recorders. Our problem is to pose questions in the right way and to search perceptively for answers from these informants. They are exposures of freshly glaciated bedrock features and surfaces and stones in glacial till, all readily available to us. Much of what is known about glacial erosion has been obtained from studies of such exposures and stones.

TYPES OF GLACIAL EROSION

Glaciers erode their beds by abrasion and plucking. **Abrasion** is simply the grinding down of a rock surface, as by a sanding machine. Close inspection of recently glaciated bedrock areas will reveal the work of abrasion through the smoothing and rounding of protuberances and the scratching and polishing of rock surfaces (Fig. 3.4).

Plucking is a process by which glaciers loosen, pick up, and remove chips, fragments, and larger blocks of rock. A glacially eroded bedrock knob usually has a smooth side, shaped by abrasion, and an abrupt, steep, ragged side created by plucking (Plate IV). A close look quickly shows that this steep side has been eroded by a process other than abrasion. Plucking works best in bedrock areas with at least modest relief and is favored by well-jointed (fractured) rocks. Besides modifying preexisting bedrock features, plucking can create topographic forms by itself.

A difference of opinion exists as to the relative roles and significance of abrasion and plucking in glacial erosion, especially in creating larger landscape forms, but few people challenge the concept that glaciers are powerful agents of erosion.

Factors influencing glacial erosion

Temperature, amount and nature of debris in the basal ice, liquid water, overburden pressure as determined by ice thickness, shear stress at the bed (which depends upon flow velocity), and the resistance and structure of bed materials are some of the factors influencing glacial erosion. With at least six independent or dependent **variables**, it is clear that glacial erosion is not a simple matter.

As to temperature, the difference between a warm glacier at the pressure melting point, therefore loose on its bed, and a cold glacier, solidly frozen to the bed, is profound. The warm glacier continuously slips over its bed; the cold glacier slips only occasionally and locally, if at all.

Clean ice is not an effective agent of abrasion, but ice laden with bits of mineral and rock particles is (Fig. 5.1). The difference is like trying to smooth a board by rubbing with an ice cube, compared to rubbing with sandpaper. The first procedure is likely to produce nothing more than meltwater. Furthermore, if we just press without rubbing, as a cold glacier mostly does, neither ice cubes nor sandpaper will be effective. We should keep in mind, however, that ice in a cold glacier above the immobile basal skin is moving, and particles anchored in it, and large

Figure 5.1. *You are in an open cavity under the Glacier d'Argentiere, in the French Alps, looking directly up at the bottom of the ice, which makes a ceiling. The linear markings extending left to right are scratches and grooves carved in the ice as it slipped over the bed. Walls of the fractures breaking through the basal ice layer show its crude lamination. This is the regelation layer, and it is relatively rich in fine rock debris. The upper black areas are coarse glacial ice, and some larger ice crystals (C) are faintly discernible. The small rounded knob (K) in the upper right probably marks a larger rock fragment within the basal ice. R is the rock bed of the glacier. This natural cavity was reached by way of a rock tunnel made under the ice by a power company to obtain water. Ice thickness is about 100 meters. (Photo by Barclay Kamb.)*

enough to project through the basal skin, can scratch the bed. Cold glaciers thus accomplish some abrasion, but not a great deal.

Water at the bed of a warm glacier inhibits more than it enhances the abrasion process, because it keeps small particles from making good contact with the bed. Liquid water, however, can be an asset to plucking. Furthermore, in concentrated flows, it can directly erode the bed on its own. Small-scale fluting, grooving, and polishing observed on some glaciated rock exposures and usually attributed to ice abrasion may actually be the product of erosion by fast-flowing streams or sheets of subglacial water. If there is enough subglacial water to form channelized streams, these flows are capable of significant erosion, especially if well supplied with debris particles. The thicker and more rapidly flowing a glacier, the less likely it is that subglacial streams can maintain fixed channels. Quasi-plastic deformation closes tunnels carved in ice more effectively under thicker ice, and movement of the glacier displaces them. Subglacial streams stable enough to carry out significant erosion are more likely under thin, sluggish, or stagnant ice near the terminus.

In Antarctica's Wright Dry Valley there is a large, spectacular, mystifying maze of interwoven steep-walled bedrock channels called The Labyrinth. Wright Valley was formerly occupied by a large ice stream, and some investigators attribute the channels to a maze of subglacial streams. Not everyone agrees, but this is a possible working hypothesis.

If we are serious about sandpapering our board, we will press down firmly as we rub. Glaciers do the same, and we would expect abrasion to be greater under thick ice. Rubbing rapidly is more effective, temporally at least, than slow rubbing, so velocity of ice flow is clearly a factor. Interestingly, however, there appears to be a limit to the beneficial effects of increases in both overburden pressure and flow velocity at the bottom of a glacier. Curves showing relationships between abrasion and increasing pressure and velocity eventually attain a peak and start to decline (Fig. 5.2).

To understand why this happens, consider a large, angular sand grain anchored in ice at the base of a glacier. To abrade effectively, it has to be solidly gripped in the moving ice and pressed firmly against the glacier's bed. Somewhere along the curves of increasing pressure and shear stress, pressure melting of ice begins in places around the grain, loosening the glacier's grasp. This lessens the amount of abrasion even though **overburden pressure** and flow velocity have increased. Because total abrasion is the work of particles of a variety of sizes and shapes, we have to be concerned with the average effect of a spectrum of responses to increased pressure and stress, but this spectrum response is believed to follow the calculated curves of Figure 5.2.

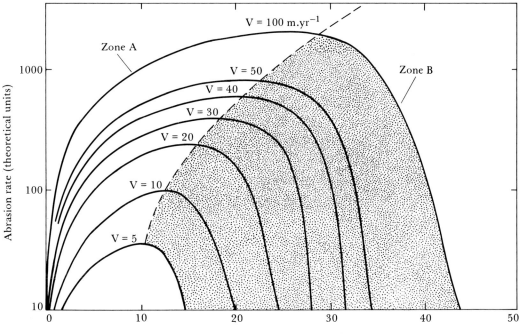

Figure 5.2. The results of theoretical calculations relating the rate of bedrock abrasion beneath a glacier to ice thickness (effective overburden pressure) and the velocity (V) of glacier flow. Figures on the vertical axis represent the abrasion rate in theoretical units; the effective over-burden pressure, plotted along the horizontal axis, is measured in bars (1 bar is the equivalent of 1 atmosphere); and the velocity curves are expressed in meters per year. The dashed line connects the peak of each curve and separates Zone A from Zone B (stippled). Abrasion increases with both pressure and velocity within Zone A, but decreases with increasing pressure and velocity within Zone B. Abrasion ceases altogether at those points where a velocity curve intersects the horizontal base of the plot in Zone B. Indeed, abrasion changes to deposition when a velocity line extends below the horizontal base. This curious behavior, more fully explained in the text, involves principally pressure melting and quasi-plastic deformation of ice holding the abrading rock particles. (From G.S. Boulton, 1974, "Processes and patterns of glacial erosion." Chapter 2 in Glacial Geomorphology, *D.R. Coates, ed., Binghamton: State University of New York, pp. 41–87.)*

The curves of Figure 5.2 merit one further comment. They predict that under greatly increased pressure and shear stress, the glacier eventually stops eroding and shifts to deposition. This presumably occurs because melting at the glacier's base, owing to increasing pressure, stress, and friction, becomes so great that debris in the basal ice is released.

It is recognized that the nature of the bedrock plays a significant role in abrasion and plucking. Quartzite, a hard, partly recrystallized rock composed of the mineral quartz (seven on the **hardness scale** of ten), is much less easily abraded than an exposure of marble composed of the mineral **calcite** (**hardness** of three). Conversely, the quartzite may be

83

much more susceptible to plucking because of its brittleness and tendency to be jointed. Marble is not particularly brittle and yields to tectonic deformation primarily by solid flow rather than by fracturing.

Glaciers, like other agents of erosion and weathering, are adept at seeking out and exploiting zones of weakness in rocks. Thus, structural planes of weakness, such as layering in sedimentary rocks, **jointing** in some lava flows, lamination in deformed metamorphic rocks, and other inhomogeneities, are exploited by eroding glaciers. For big, massive, slow-moving, seemingly insensitive bodies accustomed to riding rough-shod over obstructions, they can be remarkably subtle and sophisticated in seeking out inhomogeneities.

The permeability of the glacier bed influences the rate and nature of subglacial erosion. An impermeable rock is harder to abrade because water is concentrated at the bed, weakening the contact between it and the ice, which decreases abrasion. Because little water penetrates impervious rock, it is less susceptible to breakage by the freezing and thawing that aids plucking. The reverse applies to pervious rock: It favors both abrasion and plucking by taking water from the glacier into itself.

The abrading process

Variations in the number and nature of rock and mineral particles within the lowermost ice layer, particularly of a warm glacier (Fig. 5.1), strongly influence the degree of abrasion and account for some of the differences in features formed by it. Angular fragments of hard minerals, such as quartz and some silicates, are particularly effective abraders. Diamonds would be ideal and are found in glacial deposits, but too rarely to be significant. **Clay** is a good polishing agent, but not of much use in reducing rocks by abrasion.

Abrasion is a significant mechanism of glacial erosion. It goes on continuously, twenty-four hours a day, year after year, under a moving glacier. The glacier is hooked on the process and can't shake the habit. Abrasion may be overrated by observers, however, because of the striking aspects of its markings on freshly glaciated outcrops.

The plucking process

The small amount of water that forms by pressure melting around rock protuberances on a glacier's bed is a remarkably effective agent of erosion. If the bedrock is at all permeable, some of this water penetrates the rock's cracks and pores and upon refreezing shatters the rock. The broken fragments are then picked up and carried away by the glacier. Although this is not the only form of glacier plucking, it is probably the most significant because it goes on over wide expanses of the glacier floor continuously as the glacier moves.

The other major mode of plucking involves the removal of large, joint-defined blocks from bedrock exposures. Such blocks are usually scattered over the surface downstream from the source (Plate IV). Jointing in the rocks is important. A **joint** is simply one of a series of usually parallel planar fractures cutting through rock. Joints commonly form in sets of three separate orientations that define roughly rectangular blocks. Such blocks are probably removed largely by the push of the ice as it descends a steep face. Much glacial excavation has been accomplished this way; it is an efficient mechanism.

Fluctuations in pressure on the bed of a glacier are said in some instances to be great enough to crush, shatter, or fracture brittle bedrock, producing fragments that the glacier removes. This is another possible plucking mechanism.

Abrasion versus plucking

Differences of opinion exist among glacial geologists as to the relative amounts of subglacial erosion accomplished by abrasion and by plucking. Suppose you had been ordered to remove a large rock exposure in the middle of the front lawn and were offered a stack of sandpaper sheets or a sledgehammer and wheelbarrow to do the job. Probably most of us would choose the sledgehammer and wheelbarrow, aiming to break the rock into fragments that could be carted to the city dump. Wearing the exposure down with sandpaper looks like a lifetime task, even though disposing of the resulting fine particles could be easy. Those who selected the sledgehammer might entertain a prejudice in favor of plucking as the more effective glacial erosive process. Those favoring the sandpaper would point out that plucking, although admittedly effective, may operate only locally and intermittently compared to abrasion, which goes on continuously and automatically wherever moving, debris-laden ice is in contact with its bed.

Evaluating the relative roles of these two major modes of glacial erosion is so difficult that any person's answer is as likely to be based as much on intuition as on fact. Erosional landscape forms and the general composition of glacial deposits suggest that plucking, in terms of weight of material removed, may be superior. The greater abundance of rock chips and fragments, compared to the amount of fines, in the basal layer of glacier ice supports this view.

AMOUNT AND RATES OF GLACIAL EROSION

The amount of erosion by continental ice sheets is especially difficult to determine because the form and altitude of the preglacial surface is not normally known. Usually for reasons related to the nature and structure of the underlying bedrock or to preglacial topography, ice sheets can locally excavate closed basins of great size and depth. For example, in the heavily glaciated Mackenzie District of Canada's Northwest Territories are two huge rock basins created by glacial excavation. One is occupied by Great Slave Lake, the other by Great Bear Lake. Great Slave Lake exceeds 600 meters in depth, and its floor is more than 400 meters below sea level; the depth of Great Bear Lake is 400 meters, and its floor is also below sea level.

Because all materials laid down by continental ice sheets had to be entrained by glacial erosion, one way of getting at its magnitude is to determine the volume of glacial drift deposited. **Glacial drift** includes all materials once handled by ice. This is not easily done, because the drift occurs in small, separated, irregular patches. Even where it forms a continuous sheet, reliable figures for its average thickness are scant and highly variable. The unknown quantity of fine debris carried away by water and wind is a further complication. Nonetheless, such estimates have been attempted for the **Laurentide Ice Sheet** of North America with the conclusion that an average of only 10 meters of material, such as rock, soil, and dirt, were removed from the land surface beneath that body. This surprisingly low figure may be partly due to the low relief and hard rocks within the Canadian Shield (Fig. 5.3). (A **continental shield** is a large exposure of old, hard crystalline rocks that comprises the core of a continent.) Recently (1985) some investigators have proposed that the amount of fine glacial silt carried away by glacier-fed rivers and deposited in the ocean far exceeds the volume of drift deposited on land by continental ice sheets. On this basis they estimate that a rock layer averaging about 120 meters thick has been removed from the central area covered by the Laurentide Ice Sheet. This sounds reasonable, and this interesting idea will now be scrutinized and evaluated by glacial geologists and seafloor oceanographers.

Imagine that we have climbed a high vantage point in a rugged mountain range from which we can look down into a canyon formerly occupied by a large, powerful ice stream (Fig. 5.4), and at the same time view another nearby canyon carved solely by stream erosion. We would see that the glacier had widened, straightened, and deepened its canyon, which before glaciation looked like the V-shaped stream-cut canyon. From the U-shape of the glaciated canyon, we could determine that the

Figure 5.3. A part of central Canada scoured and plucked by the North American ice sheet. The old, hard crystalline rocks of the region are cut by fractures that clearly influenced the glacial erosional processes, especially plucking. A major fracture zone (F F F) runs obliquely across the center of the picture. The excavated rock debris was carried away to be deposited in southern Canada and the upper midwest of the United States. (Photograph ©09-09-1933 from the collection of the National Air Photo Library, Her Majesty the Queen in Right of Canada, reproduced by permission of Energy, Mines and Resources Canada, roll number A4644, frame 235L.)

widening amounted to as much as several hundred meters on each side. The degree of deepening is harder to determine, but it too was probably on the order of hundreds of meters. The floor of a part of Yosemite Valley, for example, is known from geophysical studies to have been overdeepened at least 450 meters by glaciers. The discordance of some hanging valleys created by glaciation is similar in magnitude.

Fjords deeply indenting glaciated mountainous coastlines testify to the effectiveness of ice-stream erosion. A down-at-the-heel longitudinal floor profile that gets deeper up-valley and a shallow bedrock **sill** at the mouth are characteristic. One famous Norwegian fjord has a closed bedrock basin 900 meters deep on its floor, which was shaped by glacial excavation. Water depths in fjords are impressive, commonly over 1,000 meters with a recorded maximum of 1,933 meters.

Figure 5.4. A side-looking aerial view of a glacier-filled fjord on the right and a deglaciated U-shaped valley on the left along the coast of Greenland. The flat ice masses floating in the ocean beyond are mostly, if not entirely, floes from arctic sea ice, not icebergs. (Photo by U.S. Army Air Force [10R-70] [2-54].)

To displace 1,900 meters of water and maintain contact with its bed, an ice stream would have to be more than 2,111 meters thick because of the lower density of glacial ice (0.9), compared to water. That is an impressive but not impossible thickness. A glacier nearly afloat exerts minimal pressure on its bed and erodes weakly. That may account for the down-at-the-heel profile of fjord valley floors. The sill at the mouth probably marks the location at which the ice began to float as it spread out and thinned. Flotation occurs, of course, only where the ice actually comes in contact with the ocean.

The rate of glacial erosion is influenced by so many variable factors that average figures do not have much meaning for specific situations. Subglacial abrasion is said to proceed at rates between 0.5 and 5.0 mm per year, with 1.0 mm regarded as a reasonable average. Plucking is a localized process, and extrapolation to an entire glaciated surface would not be meaningful.

Interesting experiments on subglacial abrasion have recently been performed. Plates of aluminum and marble were placed at the base of Glacier d'Argentiere in the French Alps under ice more than 100 meters thick, which is relatively thin, moving at 250 meters per year, a typical speed. The marble plate recorded 36 mm of erosion in one year. That is most impressive, but remember, marble is soft. A harder rock, basalt, placed under ice of another glacier only 15 meters thick and moving 9.6 meters per year – that's both thin and slow – recorded only 1 mm of annual erosion. That's not bad, but at 1 mm per year, 300,000 years of continuous abrasion would be required to deepen a valley by 300 meters. That seems rather slow. At 36 mm per year, only 10,000 years would be required, which seems unreasonably fast. Further experiments employing rocks other than marble under ice as thick and as fast-moving as possible would be enlightening.

Most glaciers probably abrade rocks at rates of one to several millimeters per year. Areas richly dotted with large **erratics** – stones carried long distances by ice – testify to the effectiveness of glacial plucking (Plate V). Combined, these relationships demonstrate the impressive ability of glaciers to erode solid rock.

RECAPITULATION

Glaciers erode principally by sandpapering (abrasion) and by plucking, or bodily removing rock fragments large and small. Many factors influence glacial erosion, but environments associated with warm glaciers are the most favorable. Experts are not agreed as to the effectiveness of abrasion compared to plucking, but in jointed rocks plucking is probably superior. Glacial erosion can be profound. Ice streams have widened and deepened mountain valleys by hundreds of meters, and continental ice sheets have excavated closed rock basins hundreds of meters below sea level. Many glaciated fjords have water depths in excess of 1,000 meters.

The next chapter describes an interesting array of small features of glaciated rock surfaces and the scenic, larger landscape forms created by glacial erosion.

6

Products of glacial erosion

The preceding chapter showed how an understanding of glacial erosion is handicapped by the difficulty of making direct observations at the bottom of a glacier, especially under thick, fast-moving ice where erosive processes are most effective. That subglacial erosion occurs, however, is beyond doubt, and the features and forms created are varied and striking.

SMALL-SCALE FEATURES OF GLACIATED BEDROCK

Imagine that we have fortuitously discovered a bedrock surface recently uncovered by a receding glacier (Fig. 3.4). Some of the features most likely to attract initial attention because of their unusual character are small-scale scratches, cracks, and miniature geometric forms of a size readily observable in a few square meters. We will see some of these best when down on our hands and knees, or even recumbent on the stomach, like botanists when studying tiny blossoms known as **belly flowers**.

Glacial striae

Something likely to attract immediate attention would be numerous little scratches, mostly a fraction of a millimeter deep and a few to several tens of centimeters long. On even surfaces of rock, these **striae** or **striations** are mostly straight and parallel. On uneven surfaces with small knobs, swales, ridges, steps, or troughs, many striae are curved and depart from parallelism in response to the microtopographic relief (Fig. 3.4). The direction of flow in the basal ice over such a surface was clearly influenced by its small-scale topographic configuration. This is probably more clearly

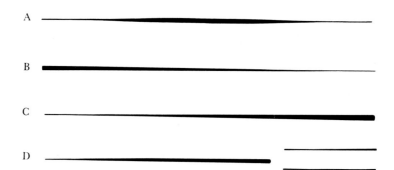

Figure 6.1. Types of striation. (A) The stria starts weakly, increases in depth and strength, and then dies out again, as though the inscribing particle had worn out or retreated into the ice. (B) A strong stria starting abruptly and then dying out as though an inscriber had rotated to bring a sharp point abruptly against the bed and then had gradually retracted or become dulled. (C) A stria starting weakly but increasing in strength to an abrupt termination, as though the inscriber had suddenly failed. (D) The most intriguing of all, a stria that terminates abruptly and is replaced by two striae continuing the same trend, as though the inscriber had broken into two pieces.

recorded by the striae than we would have been able to discern had we been at the bottom of the glacier making direct observations.

The striae have clearly been engraved by a hard, angular particle dragged across the surface under pressure and display different configurations (Fig. 6.1). Abrupt termination probably occurs because the increasing stress of carving an ever-deepening stria forces the glacier to release its grasp on the particle, allowing it to rotate to a smooth side or retract into the ice. A few striae terminate abruptly and are replaced by a double striation (Fig. 6.1D), suggesting that the inscribing particle split.

Striae are gregarious and cluster in groups; an isolated stria is a rarity. Rock exposures of only a few square meters may display hundreds (Fig. 6.2). In some settings, striae of different trends intersect. Inspection usually shows that they are of different ages; one set cuts the other. In some situations, the glacial history may indicate that crossing striae were created by ice advances separated by thousands of years. In all instances, erosion creating the younger striae was modest, or the older set would have been completely erased.

Divergences in trend, arrangement, and linearity of striae on bedrock surfaces of small-scale relief show that the ice has flowed almost like a fluid over and around minor knobs or ridges and into swales, responding sensitively to their orientation and shape. Striae tell us that warm ice under stress and high confining pressure at the base of a glacier is remarkably mobile.

In places the ice appears to have obeyed laws of fluid hydrodynamics,

Figure 6.2. Exposures like this strongly scoured, grooved, striated, and plucked bedrock exposure in the Trinity Alps of northern California leave little doubt as to the erosive power of glaciers. Ice movement was from right to left. The firm rock, a mildly altered basic igneous intrusive, records and preserves striations well. The contrast between the ragged, near-vertical plucked face (left) and the opposing scoured flank and top is impressive.

as shown by relationships depicted in Figure 6.3. This sketch is not a figment of someone's imagination; it was derived from a photograph taken in an area of recently deglaciated bedrock. Ice was flowing, as the thicker arrows indicate, downhill across a little transverse trough created by earlier glacial removal of a joint-bounded block. The trench is deepest, narrowest, and closed at the abrupt near (right) end compared to the shallower, wider, and more open far (left) end. We might suppose that this trench would simply fill up with ice, which then remains placidly in the trench as the glacier flows by. The striae show that this is not what happened. They demonstrate that ice moved out of the trench over the downstream wall and out the left end, as indicated by the thinner arrows (Fig. 6.3).

It looks as though ice was sucked from the trough by the glacier passing overhead, just as fluid is seemingly sucked out of an aspirator or spray gun by a current of air. Relationships derived from Bernoulli's

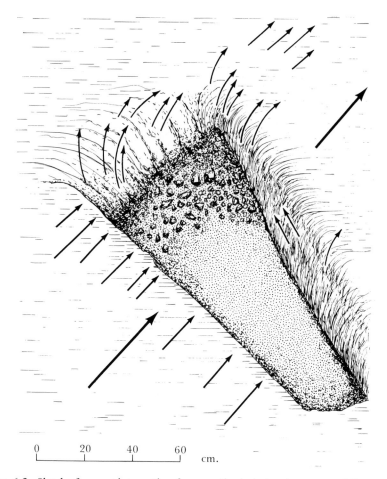

Figure 6.3. Sketch, from a photograph, of a recently deglaciated exposure of fine-grained sedimentary bedrock on the cirque floor of the former Clements Glacier near Logan Pass in Glacier National Park, Montana. The striations, light lines with barbs, show that ice lying in a small bedrock trench overridden by the flowing glacier moved up and out of the lower, more open end of the trench, obeying basic laws of fluid dynamics. Predominant direction of ice movement shown by heavy arrows. (After Max Demorest, 1938, "Ice flowage as revealed by glacial striae." Journal of Geology 46, 700–25.)

theorem explain what actually happens. **Bernoulli's theorem** says that within a homogeneous fluid a current of high velocity creates conditions of lower pressure. An aircraft wing functions on that principle. The wing's lower surface is more nearly planar and the upper surface is curved more sharply upward to create higher velocity, hence lower pressure, in the air passing on the upper side. The higher pressure underneath the wing, therefore, pushes up, producing lift that makes the airplane fly. The theorem further states that the movement within a continuous or connected body of such fluid is from high to low pressure. Ice in the little trough is slow-moving; hence, it is in a state of higher pressure than the

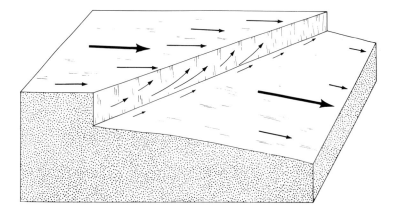

Figure 6.4. A small vertical bedrock step overridden by flowing ice in same setting as Figure 6.3. Striae, the light lines with barbs, show that ice flowed obliquely upward at a low angle along the vertical riser face of the step, and in a direction strongly oblique to the prevailing direction of glacier movement (heavy arrows). The lee-side ice was moving from an area of higher pressure to an area of lower pressure. The vertical step face is about 60 cm high. (After Max Demorest, 1938, "Ice flowage as revealed by glacial striae." Journal of Geology 46, 700–25.)

much faster-moving ice outside the trough. Consequently, ice in the trough moves up and out of the trough in response to this pressure difference, the more open left end being the easiest escape route. New ice must be continually pressed into the trough by the weight of the overlying glacier to replace the ice being removed. It does not become empty like the reservoir of an aspirator or spray gun.

Nature performed an elegant experiment in **hydrodynamics** at the small transverse trench of Figure 6.3, and the striae faithfully record the results. A somewhat similar experiment was conducted in the same area by ice flowing across a transverse bedrock step facing downstream. Striae on the vertical riser (Fig. 6.4) indicate that the ice lying against it was flowing obliquely upward and nearly at right angles to the prevailing direction of glacier movement.

The abrasion of striated stones within glacial deposits almost certainly occurred when they were at the base of a glacier, firmly held either in the ice or in the bed. Striae on stones may occur on different faces and have different trends, including a crossing relationship. This suggests that the stones shifted position and orientation. Parallel striations displayed on only one strongly abraded face by some rock fragments (Fig. 6.5) indicate that the fragment either was unusually well anchored in ice or bed or was plucked from an abraded bedrock surface.

Glacial polish (Fig. 3.4) on bedrock can be eye-catching (Fig. 6.6)

Figure 6.5. A firmly anchored limestone boulder, 75 cm across, overridden within the past century or so by Saskatchewan Glacier in Jasper National Park, Canadian Rockies. Sedimentary banding in the rock, extending left to right, is crossed by slightly radial striations. View is from the upstream, or stoss, side.

when viewed under low-angle incident light. Magnified examination of such surfaces with a hand lens reveals a huge number of little scratches essentially covering the entire area. The polished appearance is created by light reflected from the walls of this complex of scratches. Hard, resistant bedrock and an abundance of fine debris in the basal ice favor polishing. It deteriorates rapidly once the rock surface is exposed to weathering agents, and it is usually seen only on recently glaciated exposures or those recently stripped of a protective mantle.

Striae are formed most readily on soft rocks like limestone, but

*Figure 6.6. Remnant patches of glacially polished granitic bedrock in Yosemite National Park, Sierra Nevada, California, with large erratic boulders (E, background) left on the polished surface by the receding ice. Glacial polish is fragile and quickly destroyed by weathering, hence rare. Polish has been removed in the rough areas (R), at least partly by **spalling** (the flaking off of sheetlike rock fragments). Polished exposures are found mostly in recently glaciated spots or in areas protected by a recently removed mantle of glacial deposits. The polished appearance is caused by reflections from countless tiny glacial scratches on the rock surface.*

they are better preserved on hard rocks like quartzite. In glaciated areas the surfaces of quartz veins or of fine-grained aplite dikes are good places to look for striae. A **dike** is a tabular body of intrusive igneous rock, and **aplite** is a fine-grained **granite**, low in iron and magnesium.

Glacial grooves

Associated with striae, but less abundant, are larger, longer, deeper, linear, U-shaped grooves with smooth bottoms and walls and rounded edges (Fig. 6.2). Striae within **glacial grooves** are more continuous and uniformly linear than on adjacent rock surfaces. A typical groove is a few to 10 or 20 cm deep, twice as wide, and up to several tens of meters long, but they can be as much as 1 or 2 meters deep and 50 to 100 meters in length. Glacial grooves of abnormal size have been identified on aerial photographs within Mackenzie Valley of northwestern Canada. These forms, with depths to 30 meters, widths of 100 meters, and lengths to 12 km, are truly landscape features.

The smooth, worn, striated, and locally polished surfaces of grooves show that they have been shaped primarily by abrasion. Grooves form along weak zones in rock, such as fractures, **bedding**, or other layering, but many also cut directly across rock structures. They reflect an unusual concentration of the eroding power of glacier ice. Once initiated, grooves are probably self-enforcing because they channelize ice flow. A groove may begin as an unusually deep linear gouge, inscribed by a large, tough chunk of rock in the basal ice, or it may be initiated by accelerated abrasion along a high-velocity ice current. Some grooves possibly form along the trace of linear streaks of concentrated abrading debris.

Friction cracks

The next time you are served a gelatinous pudding, before plunging a spoon into it, push that spoon, bowl side down, firmly and rapidly across the congealed surface. Properly executed, this experiment creates cracks inclining steeply into the pudding and describing an arcuate form on its surface, either concave or convex in the direction of movement. We have created friction cracks, or what might more properly be termed stress cracks, friction merely being the medium that transmitted the stress generated by the moving spoon.

Hard rock is a far cry from pudding, but the combination of high overburden pressure and rapidly moving ice can generate enough stress to fracture brittle bedrock. **Friction cracks** are found on some glacially abraded rock surfaces. Trains of cracks may have been created through the focused stress generated by a good-sized stone carried at the bottom of the ice.

The most common glacial friction cracks are those called **crescentic fractures**, which usually form trains parallel to the direction of movement (Fig. 6.7). The name comes from the trace of the steeply inclined cracks on the glaciated surface. The cracks penetrate the rock for only a centimeter or two, and their trace commonly measures 10 to 20 cm, tip to tip. Curvature of individual cracks within a train is consistent, but unfortunately not all trains are consistently concave or convex downstream. We say "unfortunately" because, if curvature orientation were consistent, it would indicate the true direction of ice movement. Inclination of the fractures is commonly, but not in-variably, downstream.

Figure 6.7. Crescentic fractures on glacially polished and striated, relatively fine-grained igneous rock near the lip of Nevada Falls in Yosemite National Park, California. Such fractures, penetrating steeply into the rock for only a centimeter or two, are thought to be created by the frictional drag of fine debris in the overriding ice or by the flat surface of a larger stone carried in that ice. Ice was moving toward the observer. The polished and striated surface has been removed by postglacial weathering in the rough areas on the sides and to the rear.

In addition, a whole family of crudely crescentic depressions, known as **sickle troughs**, exists on glaciated rock surfaces. Such troughs are presumably broken out along fractures in the rock created by stress from overriding ice. They are more irregular in shape and orientation than other friction-generated features. Removal of the fracture-defined block of a sickle trough occurred while the ice was still active because the troughs are modified and smoothed by abrasion.

Figure 6.8. This view of the west wall of the canyon of the Middle Fork of the San Joaquin River, Sierra Nevada, California, shows the effects of strong glacial erosion in the form of elongate, smooth bedrock whalebacks aligned in the direction of ice flow, right to left. The rock is metamorphosed volcanics with a crude, near-vertical layering parallel to the flow direction. Effects of scour and abrasion are evident in the smoothness and streamlining of the whaleback ridges.

LARGER FEATURES OF GLACIATED BEDROCK

We have employed hands-and-knees observations to study relatively small glacial features, but now we should step back and take a more encompassing view of a glaciated bedrock terrain displaying forms that are truly landscape features. Attention centers more on topographic expression than on details of the glaciated rock surface.

Whalebacks

Whalebacks are bedrock features that occur singly but are more common in schools (Fig. 6.8). The name relates to their smoothed, somewhat elongate, gently curved form resembling the back of a whale headed for

Plate I. *This is the tidal snout of Hubbard Glacier, one of the great valley ice streams of North America. It rises deep within the St. Elias Mountains of Canada's Yukon Territory and terminates in a calving cliff at the head of Yakutat Bay on Alaska's south coast. Long-continued slow advance of this glacier culminated in 1986 with the cutoff and damming of Russell Fjord, threatening devastation of the marine wildlife of the fjord and salmon fisheries of Situk River. This was averted when the dam failed. The glacier was named not for the well-known glacier priest, Jesuit Father Hubbard, but for a member of the Canada–United States boundary survey party. (August 10, 1951.)*

Plate II. This photo beautifully illustrates the potential hazards of snow-bridged crevasses. A week or two earlier the man shown could have unwittingly walked onto the obviously treacherous snow bridging the crevasse. Scientifically, the crevasse is most useful in revealing the stratigraphy of the accumulated firn on the upper Seward Glacier in the Yukon's St. Elias Range. The man stands on the coarse, somewhat dirty snow surface of 1948; successive dirty summer layers down to 1944 are identified. Measurement of the thickness and density of each annual firn layer between dirty bands helps determine the glacier's budget for each year. (August 27, 1948.)

Plate III. Two sets of newly formed crevasses, so young they break the cover of firn mantling ice, near the outlet from the Seward Glacier accumulation basin (Fig. 1.4) within St. Elias Mountains, Canada, as they looked in August 1948. These crevasses were formed by the accelerated flow of ice leaving the basin and starting a steep descent to the Malaspina Glacier on the coastal plain. The difference in orientation of the crevasse sets suggests a change in stress probably related to a shift in direction of flow.

Plate IV. *This small rock knob along the north shore of Lake Superior in Minnesota is a mini-roche mountonnée. The right flank has been planed and smoothed by glacial abrasion; the little vertical cliff composing the left flank has been glacially plucked. Near-vertical joints in the rock were clearly a controlling factor; joint-bounded, plucked blocks lie to the left. Glacier movement was from right to left.*

Plate V. An upland meadow richly dotted with granitic erratics left by Sierra Nevada glaciers in Tioga Pass, California. Such collections of large boulders offer eloquent testimony to the excavating and transporting power of mountain ice streams.

Plate VI. A large roche moutonnée, Lembert Dome, in Tuolumne Meadows, Yosemite National Park, California, displays the asymmetry, steep lee side, and gentler stoss face typical of such features. The relatively massive granitic rock has enough jointing (fracture sets) to support effective plucking of the lee flank. If this moutonnée was initially a knob on the preglacial surface, as seems likely, it looks as though plucking has removed nearly half the original mass. Height of the moutonnée above the surrounding meadow floor is 200 meters.

Plate VII. Bridal Veil Falls, 190 meters high, in Yosemite Valley, California, pours from the hanging canyon of Bridal Veil Creek. This tributary stream of Merced River hangs 260 meters above the Merced valley floor for three reasons: During an early glacial stage the smaller Bridal Veil ice stream could not keep pace with the downcutting of the trunk glacier, westward tilting of the Sierra Nevada fault block rejuvenated the trunk stream but not the tributary, and finally, the walls of the trunk valley were cut back by its glacier, adding to the discordance.

Plate VIII. You are looking at the dirty ice face of the lateral margin of upper Steele Glacier, in the St. Elias Range, busily at work building a lateral moraine by dumping coarse, unsorted debris onto the ground along its base. When the glacier shrinks, the debris apron at the bottom of the ice cliff will become a ridge. On a warm summer day a high level of noise is created by falling or sliding masses of rock, debris-laden streams of water pouring from the glacier, and occasional muddy debris flows. (July 6, 1941.)

the briny deep after surfacing for a blow. Where small, they might be more properly called dolphin-backs or porpoise-backs.

Whalebacks are mostly 5 to 10 meters long, 3 or 4 meters wide, and a meter or two high. Sizes can be mixed within a school. They usually display modest fore and aft asymmetry, with the upstream (or stoss) end a little blunter and steeper. The surface is usually striated and often grooved and polished. Whalebacks are good places to explore for small-scale glacial markings.

Roches moutonnées

Thus far, our discussion has emphasized erosional forms primarily produced by scour and abrasion, but plucking also does its part. **Roches moutonnées** are a good example of a bedrock landscape feature formed by abrasion and plucking combined. A roche moutonnée is a prominent bedrock knob, rounded, smoothed, grooved, striated, and locally polished by abrasion on one flank, and made steep, jagged, cliffy, and irregular on the opposing flank by plucking (Plate VI).

Each flank is impressive in its own way. The smoothed, gentler flank provides the name, a *moutonnée* being the smoothly curled wig worn by barristers and judges in early European and British courts. Some people prefer an analogy to the back end of a sheep (*mouton*). The configuration of either wig or sheep is supposed to resemble the smoothed flank. It faces upstream and is thus the stoss side. Some plucking has probably occurred on this flank, but plucking there is overshadowed by abrasion. By contrast, scant signs of abrasion are found on the lee side of the knob; plucking has completely dominated. Blocks of plucked rock are strung out downstream from some roches moutonnées (Plate IV). In addition to being steeper, the plucked face is usually also higher, because the area leeward of the knob has been lowered through excavation by plucking. The lateral sides of the knob are usually smoothed, abraded, and grooved, so abrasion features may be seen over about two-thirds of its surface.

Roches moutonnées range from knobs with dimensions measurable in a few meters to major prominences 50 to 200 meters high and two or three times as big in width and length. Many are elongated in the direction of ice flow. Roches moutonnées inhabit areas in which glaciers were in a vigorously eroding mode. They favor resistant but jointed rock. Most of them probably started as a preexisting knob, created by preglacial erosion, that was modified by glacial action. Small roches moutonnées may have been created solely by glacial erosion.

In New England the rocks composing roches moutonnées have preglacial **sheeting** joints that are thought to reflect the shape and size of the

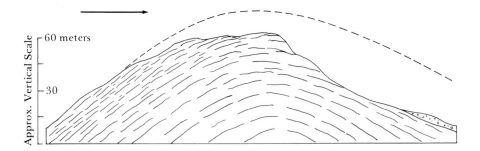

Figure 6.9. Topographic knobs within the hard crystalline rocks of New England have been modified into roches moutonnées by glacial erosion. The spacing and orientation of the gently inclined joints within the rock, known as sheeting, permit restoration (dashed line) of the original preglacial size and configuration of the knob. This permits an estimate of the amount of glacial erosion, abrasion on the stoss side and plucking on the lee side. Plucking is thought to have removed approximately ten times as much rock as abrasion. (After R.H. Jahns, 1943, "Sheet structure in granites: Its origin and use as a measure of glacial erosion in New England." Journal of Geology 51, 71–98.)

original preglacial rock knob. The form and spacing of these joints show that 3 to 4 meters of rock were removed, by plucking and abrasion, from the stoss sides of the moutonnées and that as much as 30 meters was removed from the lee sides, primarily by plucking (Fig. 6.9). This suggests that, at least locally, plucking can be ten times as effective as abrasion.

Roches moutonnées are easily recognized by their marked asymmetrical configuration, and in areas of heavily glaciated bedrock they can be a dominant feature. Watch for them: A stroll up the gentle flank can be a delightful experience; the story of their formation is there to be read.

Rock drumlins

Drumlins are hillocks molded or eroded into a streamlined form by glaciers. Many drumlins are wholly or partly of depositional origin, and the term is most commonly applied to them. Some bedrock hillocks have been eroded into a streamlined drumlinoidal form by moving ice, and they are named **rock drumlins**.

Like whalebacks, rock drumlins are smoothly abraded, but they are larger, with lengths measured in many tens to 100 meters, and their asymmetry is more marked. Rock drumlins have a steeper blunt-nose facing upstream and a narrower, gentler tail extending downstream, forming an elongated teardrop shape as viewed from above.

It is not clear why eroding glaciers form roches moutonnées in one area and elsewhere create rock drumlins. Differences in the nature and structure of the bedrock may be a factor. Roches moutonnées are mostly preexisting knobs that were modified by glacial erosion, but rock drumlins may be formed almost wholly from scratch by glacial erosion. Although plucking probably plays a major role in creating the feature destined to become a rock drumlin, the final shaping reflects abrasion. Drumlins of depositional origin are treated in Chapter 8.

Glacial basins

One thing glaciers do well, given favorable conditions such as thick, fast-moving, warm ice and well-jointed bedrock, is excavation. Localized excavation can create closed bedrock basins, many of which ultimately harbor lakes. Both ice streams and ice sheets excavate effectively, but usually in places not too close to their margins where deposition is more likely.

DIRECTION OF ICE FLOW

Many features of glacial erosion bear on this subject. In some places the direction of ice movement is obvious from geographic or topographic relationships. We know that glaciers did not form in Texas and flow north into Iowa; rather, they came south out of central Canada. Ice also generally flows downhill, and an ice stream in a mountain valley obviously descended from the high country. The direction of flow in any glacier is controlled primarily by the slope of its surface. Locally, it is possible, therefore, for ice to flow uphill if the surface slope is favorably oriented. Flow in the basalmost ice, however, is strongly influenced in detail by configuration of the bed, regardless of surface slope. Thick ice in mountainous areas can transect, that is, flow across, divides through saddles or passes. Ice divides and the topographic divides of an ice-submerged terrain do not necessarily coincide.

Striations on glaciated rock surfaces show that ice flowed in one of two opposed directions, but which one? A striation that is faint on one end and deepens progressively to an abrupt termination at the other may indicate movement toward the deep end where stress became so great that the inscribing particle broke or the ice lost its grip.

A stria that suddenly doubles may have been carved by an inscribing particle that split into two pieces, each of which subsequently made a scratch. This pattern clearly indicates the sense of movement. In places it is possible to run a finger back and forth across a glaciated surface parallel to the striations and feel a difference in smoothness. Ice was probably moving in the direction of greater smoothness. If a striated rock surface has tiny knobs of resistant material projecting a few millimeters, those knobs may protect little leeside ridges of rock, called **rattails**. They are reliable indicators of flow sense. If three-quarters of all the friction cracks in a local area suggest ice movement in a specific sense, one is inclined to heed that testimony, especially if it is corroborated by other evidence.

Among larger erosional features, roches moutonnées are excellent indicators of movement sense. Rock drumlins can be good, if their asymmetry is strong. Further attention is given this problem in Chapter 8, which deals with depositional features.

LANDSCAPES OF GLACIATED MOUNTAINS

Spectacular scenic features of heavily glaciated mountain ranges are impressive because of size, steepness, and the colossal erosion that created them. U-shaped canyons, hanging valleys, cascades, waterfalls, giant stairways, cirques, horns, and features of glacially sapped range crests combine to please the eye, heart, and mind. Books have been written about the glacial scenery of many well-known places, but a view of the actual features is more compelling than any description or photograph.

U-shaped valleys

Most valleys in mountainous terrains have been initiated, if not entirely carved, by stream erosion. In glaciated mountains, some have been modified by glacial erosion, but few if any have been created by glaciers alone. Glaciers take the easiest paths from highlands to lower areas, and those routes are usually preglacial, stream-cut canyons. The modifications glaciers impose upon such canyons, however, can be great.

Few rivers flow in straight lines, unless guided by some linear structure in the bedrock, such as stratification or a fault zone. A **fault** is a large fracture in Earth's crust along which blocks on opposite sides have moved past each other. Besides normally following an irregular course with numerous bends and turns, most rivers have tributaries on both sides. Tributary canyons on the same side of a trunk stream are separated by ridges that project as **spurs** into the trunk canyon. The trunk stream traces a course around the tips of these spurs. Viewed from a distance, spurs from opposite sides appear to overlap. Stream-cut canyons with narrow floors and steep walls have a V-shaped cross section. Only rarely are the floor of the trunk canyon and its stream visible except from directly above.

The sharp corners and narrow, irregular, confined courses created by rivers are not well-suited for streams of ice. A glacier sets about modifying a river-cut canyon by making it wider and straighter, a form more favorable for discharging large amounts of ice. **Overlapping spurs** are truncated, and valley walls are cut back and greatly steepened. Ice erodes most effectively where it is thickest, at the bottom of the valley wall. The results are like those associated with converting an old, narrow, twisting mountain road to a modern high-speed expressway. The original wandering **V-shaped canyon** is straightened and converted to an open **U-shaped valley**.

A glacier also deepens the valley, not because it needs to, but because it can't help it. In the process it may create reaches of relatively gentle gradient, beloved by hikers and backpackers. These gentle treads come at a price, however, for they are separated by steep bedrock risers a few tens to 600 meters high. The alternation of treads and risers constitutes a **giant stairway** – a succession of **glacial steps**.

Ice streams have a tendency to excavate unevenly, and in forming treads they may overdo the job, creating a succession of closed rock basins on tread surfaces. These basins ultimately contain lakes that are joined by the trunk valley stream into a chain, appropriately known as **paternoster lakes**, from the resemblance to a string of rosary beads. Lakes in glacial valleys run the risk of rapid filling with outwash debris supplied by a receding glacier. Following the recession of its last ice stream Yosemite Valley was temporarily occupied by a lake nearly 9 km long and 100 meters deep. This lake, however, filled rapidly with debris from wasting ice streams lying farther up Merced and Tenaya canyons. As a bedrock feature, Yosemite Valley is about 600 meters deeper than it appears to tourists standing on its wide, flat floor below Glacier Point or opposite Yosemite Falls.

Geologists have long puzzled over the origin of glacial steps. Clearly, they are a product of differential glacial erosion, with the brink of the riser marking the up-valley limit of a deeply eroded reach. In some instances the riser occurs on a contact between resistant, hard-to-erode rock and more easily eroded rock downstream underlying the tread. In some steps the rock composing riser and tread may be the same but different in structure; rock underlying the tread is highly jointed, but that composing the riser is massive and nearly joint-free.

Some glacial steps may reflect an original irregularity in the stream-cut profile. Perhaps a small glacial step was initiated by an icefall over an unusually steep reach in the original valley. For many reasons glacial erosion is thought to be especially efficient at icefalls. Once a small riser is initiated, it may work headward, getting higher as it goes and leaving a relatively gentle tread in its wake.

Some glacial steps are located at the junction of a major tributary. The increased ice may have led to deeper valley-floor erosion below the junction, making a step in the trunk-valley profile. In many instances, seemingly no reason exists for a step being where it is. Glaciers are capable of doing curious things. A veteran Alaskan geologist once said, "A glacier can do anything but climb a tree, and once I even saw one doing that."

Fjords have a U shape with steep walls, hanging tributaries (with waterfalls that plunge directly into the sea), and floors with steps, sills, and closed basins, just like landlocked glaciated valleys. Fjords can be very large; Scoresby Sound on the east coast of Greenland has a length of nearly 350 km and a greatest depth of 1,450 meters. Chatham Strait and the Lynn Canal, in Alaska's southeast panhandle, make a continuous fjordlike feature at least 370 km long.

Even though we may not fully understand how an ice stream creates an open, smooth-floored, glacially stepped, U-shaped valley, let us be thankful for its compelling scenic beauty and the placid restfulness of its lakes.

Hanging valleys

Few landscape features are esthetically more beautiful than a **hanging valley** endowed with a high cascade or waterfall (Plate VII).

When rivers create a drainage system consisting of a trunk with tributaries, each tributary does its best to make an **accordant junction** with the trunk stream. It tries to merge with the trunk at a common level, without a break in the longitudinal profile of either. In a **discordant junction**, one of the valley floors, nearly always the tributary, lies or hangs distinctly above the other. The difference in level can be hundreds of meters.

Hanging valleys form because of differences in the erosional resistance of rocks, because of tectonic deformation that benefits the trunk more than the tributary, or because of glaciation. No census has been compiled, but if one were made, almost surely it would show that glaciation has created more hanging valleys than any other process – perhaps more than all other processes combined.

Consider a trunk valley occupied by a large, powerful ice stream, and assume that an adjoining tributary valley is ice-free. The river in the tributary is simply not capable of cutting down as rapidly as the trunk-valley ice stream. It soon gets left behind in a hanging relationship. Even if the tributary valley has its own glacier, it is likely to be smaller than the trunk ice stream and unable to keep pace in valley deepening. In fact, neither valley need be deepened; just cutting back the walls of the trunk will create a hanging relationship.

Cirques

Most glaciated canyons have an open, theaterlike basin at their head, called a **cirque**. Cirques also occupy the flanks of high glaciated peaks, independent of any valley. In some places two or more adjacent basins have merged into a **compound cirque**, with scalloped **headwall** and outline (Fig. 6.10). Cirques are high-country features, there associated with other crestal and divide forms.

Besides a theaterlike outline, opening downslope, any cirque usually has a steep headwall and similar side walls, with a gently undulating floor sporting ponds, meadows, and rounded rock knobs. Some cirques have been so overdeepened by glacial excavation as to have a topographic closure as deep as 100 meters behind a bedrock sill. These basins usually contain sparkling, crystal-clear **cirque lakes**.

Inspection of rock exposures on cirque floors provides adequate evidence that they were once occupied by moving ice and owe their origin primarily to glacial erosion. Why erosion has been so effective at the head of a glacier, where the ice is thin and slow-moving, and how it acts to create a theaterlike basin with precipitous rock walls, are questions that furrow glaciologists' brows.

These problems are best approached by analyzing the initiation of a cirque on the flank of a lofty mountain peak. Those familiar with high country know that snowbanks commonly linger longest into the summer on the flanks of such peaks. Some snowbanks may be perennial, occupying the same spot year after year. During summer, the melting snowbank keeps the area beneath and immediately around well supplied with water, especially downslope. Water is a prime factor in weathering and erosion, and both proceed more effectively in the vicinity of snowbanks than in snow-free areas. In spring and fall, even occasionally in summer, nocturnal temperatures fall low enough to refreeze some of the daytime meltwater. Repeated freezing and thawing breaks rocks into finer particles and causes them to creep downslope. Eventually, they become so reduced in size as to be carried off by trickles of daytime meltwater.

By these means the snowbank creates a hollow in which it nestles. As

107

Figure 6.10. Vertical aerial photo of compound glacial cirques (C) in Greenland. Snow masks glaciers in the cirques and covers outward-flowing ice streams, indicated by crevasses (CR) and ice fall (IF). Near-horizontal layering in the sedimentary bedrock emphasizes the biscuit-cutter aspect of cirque morphology. A is an arête and H is a horn. (Photo by U.S. Army Air Force.)

the hollow grows, so does the snowbank, gradually increasing in thickness as well as area. Eventually, it becomes thick enough so the under part is converted to ice. When the mass attains a thickness of 60 to 100 meters, it starts to move by internal deformation and by sliding over its bed. At that point, it becomes a little glacier. Owing to movement, erosion is greatly increased, and soon the small glacier is excavating a basin for itself in the side of the peak.

Accumulation of the transported debris along the downslope edge of the ice builds an embankment, which enhances the basinlike form of the depression. It is now well on its way to becoming a cirque, and in time the little ice mass becomes a **cirque glacier**. Under some conditions it may never extend beyond the confines of its own basin, but under favorable circumstances it can grow large enough to flow out of the basin and become an ice stream.

Cirques heading glaciated canyons probably evolve in much the same way but are commonly larger. This happens because of the greater amount of snow accumulating in the protected head of a valley than on the exposed flanks of a peak. The erosion that creates canyon-head cirques may be relatively great, simply because ice has been at the task longer there than in any other part of the glacier's course. The thin ice around the sides and head of a cirque glacier favors breakup of rocks by the freeze-thaw process, which is supported by the strong solar radiation and large diurnal and seasonal temperature changes of high altitudes. Once rocks are broken and in contact with the ice, the glacier has little trouble carrying them away. The cirque glacier literally eats backward into the bounding slopes, undercutting them so they become oversteepened and contribute to the enlarging process by shedding coarse debris onto the glacier, which efficiently exports it.

This headward process is called **sapping** by geologists and engineers. As the cirque glacier eats its way almost horizontally into the mountainside, it creates the steep wall and relatively flat floor that characterize a cirque. We would expect a cirque with a broad floor to have higher walls than one with a narrow floor, and in general this seems to be true. Some of the wall height of any cirque may result from deepening, but much of it simply reflects the amount of back-cutting. One of the world's largest cirques has a floor diameter of about 16 km and a wall height of 3 km.

Features of glaciated mountain crests

Suppose two cirques are sapping headward toward each other on opposite sides of a divide. In time they will convert it to a narrow, knife-edge ridge bounded on either side by the steep headwalls of the opposed cirques. The French word **arête** is applied to such a feature. If the bedrock is homogeneous, the skyline profile of the arête may be smooth, but if the rock is heterogeneous and irregular in structure, particularly if it is jointed, the arête's profile will be rough and jagged – in a word, a **sawtoothed ridge**. Spires and joint-defined columns of rock standing on its crest are called **gendarmes** after the erect, straight-backed French

Figure 6.11. The sharp, rugged point on the left skyline is Pilot Peak, one of the finest glacial horns in North America. It has been created by headward cirque (C) excavation from several sides, dissecting the volcanic rocks of the Absaroka Range in northwestern Wyoming, just east of Yellowstone Park. This scenic, 3,415-meter peak guided explorers and trappers, especially John Colter, prowling this region in the early nineteenth century. In its shape and ruggedness it favors the famous Matterhorn of the Pennine Alps. In the midforeground is an excellent bouldery lateral moraine (M). (Photo by the Jack Richard Photo Studio, Cody, Wyoming.)

policemen. Arêtes can also be created by lateral erosion along a ridge separating two parallel ice streams.

Besides being narrowed, the divide between two opposed cirques may be lowered by hundreds of meters, forming a saddle-shaped gap, or **col**. Ice over a mountainous terrain can also become so thick in places that it buries a part of a topographic divide, forming a **transection glacier** that crosses the divide and creates a col by erosion.

Picture three or even four cirques on the flanks of a high peak all working headward toward its summit. In time the peak, regardless of its original shape, will become a steep-sided, pyramidal spire rising to a sharp point

Figure 6.12. Trimlines are graphic recorders of glacial history. When ice invades a densely forested region and then recedes, it leaves a sharp trimline between devastated and forested areas, as shown in this picture of Franklin Glacier and its tributaries in the Coast Mountains of British Columbia. Lack of significant vegetative regrowth inside the trimlines indicates that the glaciers of this region were more than 100 meters thicker and more extensive within the past 100 years. (U.S. Geological Survey photo by Austin Post, September 30, 1960.)

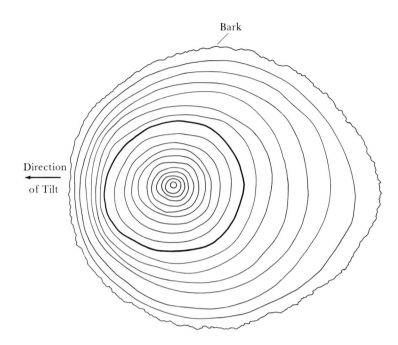

Figure 6.13. Trees growing along the edge of a trimline, tilted but not killed by the ice, form eccentric growth rings resembling those shown in this sketch, as seen in a cut across the trunk. If the tree is still living, an investigator can count years backward from the bark to determine when the tree was tilted. This fixes the approximate date of the glacier's greatest expansion.

(Fig. 6.11). This feature is a **horn**, taking its name from the famed Matterhorn of the Pennine Alps on the Swiss–Italian border near the village of Zermatt.

Trimlines

No matter how glacial erosion features are categorized, a few items always seem to be left over. One of these is treated here.

When an ice stream moves down a forested valley and then recedes, it leaves a knife-sharp line between barren deforested ground and standing timber. This is a **trimline** (Fig. 6.12). In time, new trees will grow in the deforested area, but the difference in tree size is so obvious that the line remains well defined. A line marked by trees of different age might be termed a **paleotrimline**. The oldest trees growing inside the trimline provide a minimal date for the glacier advance. An exact date could be obtained by finding a living tree along the trimline that was tilted but not killed by the advance. The rings of tilted tree trunks grow eccentrically, as shown in Figure 6.13. It is a simple matter to count the eccentric rings back to the time of tilting and thereby date the glacial advance.

RECAPITULATION

Direct evidence of glacial erosion on exposures of bedrock is expressed mostly in the form of striae, grooves, smoothing, rounding, and sharp truncation of internal rock structures. Larger features, such as U-shaped and hanging valleys, glacial steps, whalebacks, rock drumlins, roches moutonnées, excavated lake basins, cirques, and horns contribute to the scenic appeal of glaciated mountainous landscapes. Abrasion has created most of the smaller features, and plucking has usually dominated in forming the larger landscape features. Undercutting of steep slopes by glacial sapping, largely through plucking, plays a major role in the formation of cirques and other forms along high crestal divides. Some surface markings and erosional landforms, especially asymmetrical roches moutonnées, indicate the direction of ice movement.

The material derived from glacial erosion is transported and deposited by the ice. The resulting accumulations can be highly distinctive and are abundant in some populated areas. The important mechanisms of glacial transport and deposition are treated in the following chapter.

7

Glacial transport and deposition

The time-honored adage that what goes up must come down loses some of its verity each time another object is lofted into outer space. The old saying still holds, however, for glaciers. Every object they pick up eventually has to be put down, and that can amount to a lot. Glaciers are like children; they almost never put something back where they got it. They carry it off and leave it somewhere else. In a word, they are transporters. The picking-up process has been discussed; let us now consider the subjects of transport and deposition.

GLACIAL TRANSPORT

It is helpful to recall the three glacial environments previously mentioned: **superglacial** (on the surface of the ice), **englacial** (within the ice), and **subglacial** (beneath the ice). We also need to add a fourth setting, **extraglacial**, encompassing the area alongside and beyond the ice. It is further useful to distinguish two subdivisions within the extraglacial realm, peripheral and proglacial. **Peripheral** refers to the zone immediately along the edge of the ice, and **proglacial** designates the area outward for a much greater distance, generally in the direction of ice movement.

It is natural to entertain an image of a glacier acting as a gigantic bulldozer advancing across the landscape, scraping up everything in its path, and pushing a great jumbled pile of debris ahead. Most glaciers actually transport little material in a bulldozer fashion. In some places, small ridges of debris obviously pushed up by the advancing ice front of an existing glacier are seen, and in special situations the push of an advancing glacier upon previously deposited materials has created embankments up to 100 meters high. In neither instance has material been transported very far. Most glacial scraping has occurred under the ice, and not in bulldozer fashion at its edge.

Subglacial transport

Transport of rock debris underneath a glacier can occur only by **traction**. That is, the particles are not encased in the ice and are moved as individual pieces across the bed by slipping or rolling under propulsion from the overriding glacier. Observers of the subglacial realm have reported some evidence of transport in this manner. Sheets or streams of water running over the subglacial bed carry debris and should probably be included among the agents of subglacial transport. If a sheet of glacial drift beneath a glacier actually flows under the shear stress exerted by the ice, as some workers postulate, that is a special and possibly important type of subglacial transport by traction. The contribution by all forms of subglacial transport to the total work done by glaciers appears to be relatively small. Fine material suspended in water may be an exception, for subglacial water is commonly muddy (Fig. 2.18).

Englacial transport

Most debris moved by glaciers travels englacially in suspension, much of it initially in the basal and marginal parts of an ice stream. The ability of any medium – air, water, or ice – to transport particles in suspension depends upon relative densities, the medium's dynamic viscosity, and for air and water the degree of turbulence generated by flow. The higher the viscosity, the greater the power for transport in suspension. Water at 20° C has a viscosity of 0.01, as measured by a complex unit called a **poise**. Ice at 0° C has a viscosity of about $n \times 10^{14}$ poises; that means any number from one to nine followed by fourteen zeros. In terms of viscosity ice and water are simply not in the same league.

Water and air have one advantage over ice, however, because both experience strong turbulence when flowing with any appreciable velocity. The upward-directed component of turbulent flow keeps particles in suspension, like stirring a can of paint to get the color particles uniformly distributed. For reasons earlier explained, glaciers do not experience turbulence, but they can handle huge quantities of just about anything in suspension without turbulence, thanks to the high viscosity of ice.

Figure 7.1. Basal ice exposed in this terminal cliff, nearly 100 meters high, of Crillon Glacier on the southeast coast of Alaska near Lituya Bay, displays dark bands of rocky debris sheared or squeezed up from the glacier's bed, one of the mechanisms for bringing subglacial debris into an englacial position. Depending upon the relative rates of top and bottom melting, much of this debris may ultimately arrive on the glacier's surface, becoming part of the superglacial ablation mantle. (Photo by Bradford Washburn.)

Even extremely large blocks of rock, which sink into the ice owing to pressure melting and quasi-plastic flow, sink so slowly that they are transported great distances in suspension.

A glacier accumulates englacial debris within various parts of its body in several ways. The accumulation areas of many glaciers, especially ice streams, are bordered by steep, rocky slopes that shed debris by slides and rock falls onto the snow surface; then more snow accumulates on top, making the debris englacial. Debris that falls or is carried into a crevasse or moulin instantly becomes englacial.

Planar bands of rock debris (Fig. 7.1) appear to have been carried or squeezed up into the basal part of some glaciers. This probably happens in reaches of strong compressing flow, such as near the terminus or at the base of an icefall. Icefalls, with large, open crevasses and exposures of rock on their face, deliver a mixture of ice, snow, and rock fragments to the plunge pool below, where much of the debris becomes englacial. Folds in ice, probably caused by compressing flow, can raise basal debris to an englacial position.

A rock knob projecting above the glacier floor, whether it extends to the surface or not, sheds debris into the englacial realm. Where two ice streams from adjacent valleys merge to create a medial moraine, the septum of dirty ice underlying that moraine (Fig. 2.5) becomes englacial. Because of their topographic setting and makeup, ice streams generally have more englacial debris than ice sheets. Ice sheets with numerous **nunataks**, islands of rock completely surrounded by glacial ice, will have more englacial material than those without them.

Superglacial transport

Most superglacial debris within the ablation zone was formerly englacial. It became superglacial through lowering of the ice surface by melting and accordingly is called **ablation debris**.

Any loose superglacial debris has considerable mobility all its own, independent of ice movement. As the mantle becomes thicker, a rough, irregular topography develops on the ice surface by differential melting (Figs. 1.10 and 7.2). Local areas of thick debris protect the underlying ice from melting as rapidly as nearby bare or thinly mantled ice. Protected spots become high-standing knobs or ridges, but this eminence is their doom because soon debris creeps and slides off into adjoining low spots. Eventually, the high places are denuded and melt more rapidly than the adjacent debris-filled low spots, which are destined to become future high-standing features.

This inversion of topographic relief goes on continuously. The debris gets worn, broken up, even crudely sorted, during the process of being shifted back and forth from one spot to another. Streams of water and ponds on the ice surface aid in the movement, wear, sorting, and accumulation of superglacial debris (Fig. 7.3).

Huge blocks of rock, as big as a small house (Fig. 7.4), can be carried in superglacial position for great distances with ease. Such a block will remain on the ice surface rather than sink into the ice because it provides enough protection from radiative and meteorological melting to balance the pressure melting and quasi-plastic flow caused by its weight. In some settings rocks of proper size and shape provide more protection than needed, so the surrounding bare or only weakly insulated ice melts more rapidly, leaving the blocks perched on ice pillars, forming glacier tables (Fig. 1.11).

Figure 7.2. A 1941 view of striking and unusual topography on the surface of stagnant Steele Glacier. The window in the dirty ice cliff may be the remnant of an englacial tunnel. This ice is dirty and complex in structure because it has been involved in one or more major surge events. The coarse, angular character of the surficial debris characterizes much, but not all, ablation morainal material. One sees at the top of the cliff that the debris mantle can be surprisingly thin, just a skin on the ice.

Extraglacial transport

Movement of material into the peripheral subzone of the extraglacial area is by gravity, debris flowage, and running water. Gravity causes individual rock fragments to fall, roll, or slide off the ice onto adjacent ground. **Debris flowage** involves semifluid masses of water-rich debris, resembling freshly mixed concrete, flowing downslope off the glacier. Running water and wind carry much debris in their usual manner into the peripheral area and beyond. Debris flows do not usually act beyond the proglacial area, but water and wind are capable of spreading debris far beyond the immediate glacial realm.

Figure 7.3. One of the superglacial ponds shown in Figure 1.10. Debris constantly falling into the pond from the undercut receding ice face (I) keeps the water muddy during summer. A former high-water line (A) is visible midway up the ice face, and two lower lines (B and C) are preserved in the debris-mantled slope to the right. The less than 1-meter thickness of the superglacial debris mantle (arrows) is clearly visible, and the ice obviously contains adequate englacial debris to increase the thickness of that mantle as melting continues. This ice is unusually dirty because of an earlier event of glacier surging which reactivated a previously stagnant ice stream that, as in the condition pictured here, was heavily mantled by debris. This 1941 scene was completely destroyed by another surge in 1965–6.

HOW GLACIERS DEPOSIT

We will now examine differences in the mechanisms of deposition within the four environments of transport.

The load factor

Rivers commonly become overloaded, but they adjust quickly and easily by depositing excess material, usually the coarsest debris first. Glaciers

Figure 7.4. A house-sized erratic resting on the surface of the outer stagnant margin of the Malaspina piedmont ice sheet (Fig. 2.3) on Alaska's south coastal plain. This huge block was transported not less than 50 km and perhaps two or three times that distance from an unidentified source in the St. Elias Mountains to the north. It remains on the ice surface because the downward melting of the more thinly mantled surrounding ice is about the same as the pressure melting and quasi-plastic yielding of ice beneath the block. The block might once have been englacial, but it is now unquestionably supraglacial. Glaciers have a tremendous capacity for transportation. (July 6, 1948.)

cannot adjust so readily, because it is difficult for them to deposit debris quickly. The transporting power of ice is so great, however, that almost never does an active glacier become overloaded. The supposition that glaciers stop moving because they become overloaded with debris is probably pure fantasy. Heavily loaded glaciers usually contain no more than 10 percent debris, and nearly 70 percent may be required to immobilize them. The high debris:ice ratio existing in some masses of stagnant ice probably developed after motion ceased for some other reason.

Subglacial deposition

The relatively high content of debris within the basal 15 to 100 cm of a glacier is well documented (Fig. 5.1). Let us examine some factors that could be expected to cause debris to be deposited along the glacier's bed.

Warm glaciers are continuously melting at the base, owing to heat conducted from below. Under most glaciers, this heat flux is reasonably uniform. Exceptions exist in volcanically active areas like Iceland, Mount St. Helens, or Yellowstone, where glaciers do strange things in the way of transport and deposition because of hot spots on their beds, as will be shown later. Volcanism and glaciation are uneasy neighbors.

The heat generated by friction during basal slip exceeds the terrestrial heat flux and plays a major role in causing deposition. It tends to be variable in distribution and magnitude owing to topographic irregularities on the bed, differences in bed permeability, and the inhomogeneous debris content of the ice. Any topographic irregularity increases the friction generated by basal sliding and thus increases melting and the deposition of debris.

Variations in water along the bed can significantly influence deposition. Water varies widely in amount and distribution for several reasons, bed **permeability** being a primary factor. Deposition occurs over permeable spots more readily, because the paucity of water causes increased friction, greater melting, and stronger drag on the overriding ice. Debris-rich ice also generates more friction than does clean ice. These and other variables cause basal deposition to occur irregularly in time and place. In fact, both erosion and deposition can be taking place simultaneously and in proximity under a glacier.

As particles encased in basal ice are freed by melting, some are rolled or slid along the glacier's bed, and others are reclaimed by the ice through pressure melting and refreezing (regelation). Those not recaptured become lodged in some favorable spot where other particles have probably already accumulated. The term **lodgment till** is applied to these accumulations formed under moving ice. Under thin ice near glacier margins, reincorporation of particles is less, and subglacial lodgment increases.

Lodgment deposits, having accumulated under pressure, stress, and minimal free water, tend to be firm, compact, and rich in fine particles. They may display a preferred orientation of larger rock fragments of certain shapes, a phenomenon called **fabric**, which can indicate ice movement direction. Distinctly elongated stones, which are otherwise roughly equidimensional (a fat cigar shape), can be oriented with the long dimension perpendicular to ice flow. They are thought to have been rolled along the glacier's bed. Platy rock fragments in lodgment till can be oriented with flat surfaces nearly parallel to the bed except for a slight upstream tilt. These stones are thought to have moved by sliding. The upstream tilt prevents the leading edge from digging into the bed and gives the glacier a face on which to shove.

Not much subglacial deposition seems to occur in the source areas of glaciers, but some takes place almost everywhere else, especially if the floor is topographically rough. Subglacial deposition can become a dominant process under the outermost reaches of glaciers, especially ice sheets.

Superglacial deposition

Englacial debris cannot be deposited without entering the subglacial, superglacial, or extraglacial realm, so we will skip ahead to superglacial deposition. Most englacial material becomes superglacial, and its accumulation in that setting has a familiar analogue in the residual winter snowbank that appears to get dirtier and dirtier as the spring progresses. Some of that dirt settles out of the air onto the bank, but much of it represents material caught up in the snow as it accumulated or was scraped into a pile. With melting, more and more of this included dirt collects on the surface, leading to more absorption of solar radiation and faster melting. The same thing happens to glaciers. By the time the ice of a glacier is gone, all its englacial debris, except that captured by subglacial lodgment, has entered the superglacial realm.

Superglacial debris suffers one of two fates. Either it is moved off the glacier by gravity, debris flowage, or running water into the extraglacial area, or it is let down onto the ground, in situ, as the ice melts. This occurs primarily where ice has stagnated. The outer reaches of some glaciers become stagnant, because the glacier has overextended itself by short-lived surging or because unfavorable climatic change has drastically reduced its sustenance. Usually the ground under stagnant ice has previously received a thin mantle of subglacial lodgment debris, so the superglacial material comes to rest atop the lodgment layer. The two layers differ greatly in character.

The superglacial mantle undergoes so much reworking and modification before being deposited that it may look more like a poor grade of gravel than glacial till. In constitution it ranges from jumbled masses of mixed large and small angular fragments (Fig. 7.2) to localized accumulations of well-sorted and rounded stream gravels (Fig. 1.9). Fine debris is usually sparse, except where concentrated into superglacial pond deposits (Fig. 7.3). This jumbled, unconsolidated mass contrasts sharply with the tight, compact, fines-rich lodgment deposits, the stones of which may show wear by glacial abrasion, a rarity among superglacial stones.

Extraglacial deposition

Some of the material moved off a glacier accumulates alongside the ice, that is, peripherally, and some may be carried farther away into the proglacial subzone. We will consider first the accumulation of material in the peripheral subzone.

PERIPHERAL DEPOSITION.   Because ice near the outer margin of almost any glacier has inevitably undergone much melting, its surface normally bears a considerable mantle of ablation debris. Assume that the ice edge is essentially stabilized, occupying about the same position year after year except for minor seasonal fluctuations in position. In this state, the rate of forward movement is balanced by recession of the ice edge through melting. The superglacial debris is riding slowly forward, as though it were on a gigantic conveyor belt, and being dumped off the glacier to form a peripheral deposit, a **dump moraine** (Plate VIII). Given enough time, the accumulation can build up to a level where it partly buries the edge of the glacier. When the glacier recedes, this accumulation is left as a ridge outlining the formerly stabilized ice edge. Any ice buried within the debris eventually melts, contributing to the irregular topography.

If the ice edge is either advancing or receding, the conveyor belt deposits the superglacial debris in the form of a sheet, much as a highway paving machine lays down a ribbon of concrete or asphalt. The glacier is much larger and spreads over a wider front than any manufactured equipment, but the results are similar. Deposits spread by glaciers are more irregular because of the heterogeneous material and because few glacier margins move with the uniform, steady pace of a paving machine.

The dumping process encompasses more than just the falling, rolling, or sliding of rock fragments off the ice into the peripheral area. Melting is great at the outer margin of a glacier during summer, so water is abundant. Streams pouring off the glacier build little **alluvial cones** within the peripheral embankment. An alluvial cone is a hemicone-shaped accumulation of **alluvium**, which is sand and gravel carried and laid down by a stream. Where associated with glaciers, such materials are termed **glaciofluvial** to indicate their hybrid origin. If the superglacial debris is abundant and contains at least a modest amount of sand and **silt**, mixing with water can form a fluid slurry that readily flows as a mass from the ice into the peripheral subzone. This is the phenomenon of debris flowage, and the lobate tongue of material created is a **debris-flow deposit**. This is a widespread phenomenon, not restricted to glaciers. The accumulation of heterogeneous debris built up by these means is, in a sense, one of nature's garbage dumps.

PROGLACIAL DEPOSITION.   Transition from the peripheral to the proglacial environment involves a change in depositional mode with glaciofluvial processes becoming dominant. Glaciers are still active participants, because they furnish the water and debris. Both debris flowage and glaciofluvial transport carry material beyond the peripheral area into the proglacial subzone, although debris flow deposits there are less abundant. Glaciofluvial material can be carried many kilometers beyond the edge of the ice and spread in a wide sheet over the land as an **outwash plain** or deposited as a long ribbon within the confined course of a valley, making a **valley train** (Fig. 7.5).

In summer, the amount of water draining from a glacier into and across the extraglacial area can be enormous, and most of it is heavily charged with rock fragments of all sizes. The streams are rushing torrents, essentially in flood, and they rapidly perform prodigious feats of deposition and erosion. All are at least turbid, if not downright muddy, because of the abundant fine rock particles, called **rock flour**, produced by glacial abrasion (Fig. 2.18). Rock flour suspended in water supposedly makes **glacier milk**. In most instances it is not very appetizing milk, being gray, tan, or muddy because of the colors of the rocks that have been ground up.

The resulting deposits of mixed sands and gravels with crude bedding, only modest **sorting**, and many **scour-and-fill** structures, are named **outwash**, a reasonable designation for material washed out from a glacier. Poor sorting means that particles of many sizes are mixed together. Scour-and-fill structures are channels cut into earlier deposits that are refilled by younger gravels.

The steep ice walls of stream channels cut into glaciers are prone to collapse, filling the streams with icebergs. These are carried downstream, and some are buried in outwash deposits. When such an ice block melts, it creates a closed depression that may fill with water to form a pond.

In such extraglacial ponds and lakes, **glaciolacustrine** deposits accumulate, mostly fine-grained, but they can be locally coarse for the following reasons. If a lake laps against an ice face, coarse debris can fall directly into it. Ice cliffs undercut by the lake will shed bergs that drift across the lake, and debouching streams may also bring bergs. Such bergs commonly contain rock debris, some of it coarse, which is deposited on the lake floor where they melt. This is **ice-rafted** debris, and individual rocks are **dropstones**. By these means, pods of coarse material can be irregularly scattered through fine lake-bed deposits. Furthermore, any glacier-fed stream flowing into an extraglacial lake builds a delta, mostly of sand and gravel.

Finer sands and silts accumulate on the lake floor beyond the delta, where they make a soupy mixture with water. This mix accumulates to

Figure 7.5. Looking down onto the eastward extending active glaciofluvial valley train (VT) of Saskatchewan Glacier in Jasper National Park, Canadian Rockies. Abandoned braided stream channels (B) are a typical surface marking of outwash deposits. Part of the smooth, receding snout of the glacier lies in the lower right. The Saskatchewan River is fed by streams emerging from subglacial ice tunnels near the right and left sides of the ice tongue. A remnant area of ridged ground moraine (R) is faintly visible near center. (Summer 1954.)

a point where it becomes unstable and moves out across the lake bottom. As it moves, it becomes more thoroughly mixed with water and takes on the character of an underflow, or **turbidity current**. Because of high debris content, the turbidity current has a greater density than the surrounding water. It flows under gravity even on very gentle slopes and spreads out into a sheet covering much of the lake floor. There it settles and deposits a thin layer of silt and fine sand. Repeated turbidity flows generated during summer, when incoming streams carry great quantities of debris, can build up successive thin layers of silt or fine sand, to a centimeter or two in cumulative thickness, over much of the lake bottom.

With onset of winter, outwash streams entering lakes shrink dramatically and may even cease to flow; the lake freezes over and becomes thermally stabilized. Turbidity current activity ceases owing to a lack of debris. Glacier-fed streams are muddy, so extraglacial lakes receiving such waters are also muddy. Some of this fine material settles out in summer along with the coarser debris, but in winter it is the only material that continues to be deposited on the lake bed. Thus, a thin layer of the finest particles, usually of clay size (0.004 mm or smaller), forms on top of the thicker and usually lighter-colored silt-sand layer deposited by summertime turbidity currents. This pair of layers comprises a **couplet** known as a **varve** (Fig. 7.6). Each varve represents a year's deposit, and they have been called the tree rings of Earth.

Another extraglacial agent, active over outwash plains and worthy of attention, is wind. The extraglacial environment is windy for two reasons: Glaciers mostly favor stormy, windy areas, and glaciers also make their own winds. The glacier-generated winds range widely in strength depending upon size and slope of the ice mass, topographic setting, and independent meteorological factors. They can be powerful and usually prevail day after day, unless temporarily overwhelmed by stronger meteorological influences – a frontal storm, for example. They are generated as follows.

Cold air is heavier than warm air. Air in a heated room is coolest near the floor and warmest near the ceiling because warm air rises and cold air sinks. Air lying in contact with ice becomes chilled to temperatures near 0° C. In that state, it is considerably heavier than the warm air over ground off the edge of the glacier. Because cold air naturally flows downslope by gravity and the surface of most glaciers slopes outward toward the edge, the proglacial area is swept by cold winds off the glacier. These air currents are called **gravity winds**. They are not unique to glaciers, being common in hilly or mountainous terrains where cold air from high areas frequently drains down slopes and valleys. Descending air warms up because of compression, but the gravity winds of glaciers do not suffer too much on that score, because they are constantly being cooled by the underlying ice.

Any wind blowing across an area of fresh outwash finds a rich supply of sand, silt, and dust to pick up and transport. The dust and finest silt go into suspension and can be carried far away, but the **sand** has to travel by **saltation** (hopping) and **traction** (movement along the

Figure 7.6. This is a typical exposure of varved glacial lake deposits. The thinner dark layers, a fraction of a centimeter thick, are fine-grained, clay-rich winter deposits, whereas the lighter-colored layers, up to several centimeters thick, are silty or fine sandy summer deposits. Each pair of layers, a couplet, represents one year of deposition. The layers in this exposure accumulated in about twenty years. Varves are useful for dating young, relatively short-lived geological events, involving centuries or a few millennia. A cumulative varve chronology covering nearly 17,000 years has been constructed for Scandinavia.

ground). Saltation is a major mode of eolian transport for sand grains. Perhaps as much as three-fourths of all windblown sand has been moved by that means. For a sand grain, the difference between saltation and traction resembles the difference between walking and crawling for a

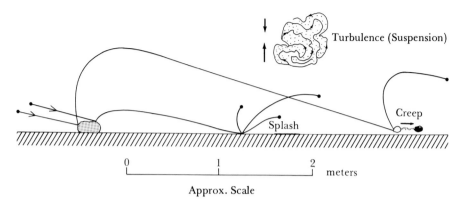

Figure 7.7. The mechanisms of suspension and saltation, accompanied by splash and impact creep, for particles being transported by wind are illustrated. The inset shows how buoyancy created by the upward component (longer, upward arrow) of complex air turbulence balances or exceeds the downward pull of gravity (shorter, downward arrow) on particles of small size or low density. Particles of sand size moving by saltation can attain velocities approaching that of the prevailing wind (easily 30 km per hour), reach heights of 1 to 2 meters, and cover distances of many meters on each hop under ideal conditions.

child. Parents can testify that a walking child covers far more ground and gets into many times more mischief.

Saltating sand grains have another less commonly recognized effect upon materials of a surface across which they travel. They move particles normally much too large to be set into motion by force of the wind alone. Most saltating grains return to the surface at a low angle (10 to 15 degrees) and with velocities around 25 km per hour. Even though they may be only 0.3 mm in diameter, this velocity imparts considerable momentum. As a grain strikes the ground, it may blast other grains into the air so they can saltate, it may loosen and loft fine dust and silt particles so they can travel in suspension, it may rebound off a larger stone making another hop, or it may impact grains 0.5 to 1.5 cm in size. The impact causes such grains to move across the ground for a centimeter or two. Were we to watch one of these grains under a heavy rain of saltating particles, it would appear to be creeping across the ground in a series of jerky advances. This is the process of **impact creep** (Fig. 7.7). In effect, the process selects particles of a specific size, transports them slowly across the ground, and eventually accumulates them in protected spots or in wavelike ridges resembling large ripples (Fig. 7.8).

Commercial sandblasting, employing a current of compressed air carrying abundant sand grains, is an efficient means of cleaning up soot-stained buildings. Sand grains moving across outwash plains by saltation strike against surface stones. This natural **sandblasting** goes on day after day for years, decades, even centuries. As a result, the stones are eroded,

Figure 7.8. These ridges, resembling large ripples, are made up of rock and mineral particles about 10 mm in size. They were driven across the ground by the impact of saltating sand grains (Fig. 7.7) a fraction of a millimeter in diameter under the influence of strong winds from the right, as shown by the asymmetry of the ridges with steep faces to the left. Particles 4 to 10 mm in size are termed granules, so these forms are called granule ripples. Each ridge represents a sort of traffic jam in which the impact-driven particles accumulate. The amplitude, or height, of such granule ripples can be as much as 10 cm. The locality is Coachella Valley, California.

pitted, etched, fluted, grooved, polished, and faceted. In some instances, more than half of a large stone can be worn away. Wind-blasted stones are **ventifacts** (windmade), and they are abundant on old, abandoned outwash deposits the world over, as well as on other barren, stony, windswept surfaces with adequate sand (Fig. 7.9).

The sand and dust picked up by wind from outwash surfaces usually end up in different depositories. The sand travels perhaps a few to several tens of kilometers to areas peripheral to the outwash plain, where it accumulates as dunes. Dust traveling by **suspension** goes much farther, hundreds to thousands of kilometers, before it settles to earth, becoming part of the soil or adding to seafloor sediments. If it accumulates predominantly in one area on land, it can form a deposit of **loess**. Not all loess is glacial in origin; some comes from deserts and dust bowls.

Figure 7.9. Sandblasting can be a very effective erosive process as this boulder on the surface of a gravel mantled incised terrace east of the Bighorn Mountains in Wyoming, near Buffalo, testifies. The deep grooving on its surface was created by saltating sand and finer particles driven by a strong wind from the right. This is a ventifact, but it is old. The cutting occurred tens of thousands of years ago, when the gravel surface was barren and at stream level.

RECAPITULATION

Glaciers have almost unlimited capacity to transport rock debris. No particle is too small, no block too big, and nearly all are carried in suspension, within or upon the ice. The principal drawback to glacial transport is the leisurely pace, averaging something less than 1 meter per day. Associated agents of transport taking over at the glacier's margin, such as debris flowage, running water, and wind, operate at a much faster pace.

Everything glaciers transport must ultimately be laid down. Some deposition occurs beneath the ice, but much transported debris is either dumped along the glacier's edge or carried away by associated processes.

Wind and water carry the finer debris, sand, silt, and dust, far beyond the glacier.

Glaciers have been able to construct some fascinating deposits and landforms out of the jumbled mess of debris they transport. We will now examine such products.

8

Products of glacial deposition

Features of glacial erosion are scenically more appealing than the products of deposition, but glacial deposits are more widespread, covering 8 percent of the world's land and nearly 25 percent of the North American continent. They are also more accessible to most people, although not always easily recognizable because of undistinguished topographic expression. Economically it is no contest: Deposition is clearly the winner. Freshly ground rock debris provided by glaciers furnishes mineral-rich parent material for much farmland, and glacier-derived loess weathers to some of the world's richest soils. Sands and gravels of glacial origin are widely exploited for building and road materials. The major commercial product arising from glacial erosion is electrical power derived from the waterfalls or cascades of hanging valleys or glacial steps. Lakes, the product of both glacial erosion and deposition, provide water for power generation, irrigation, or domestic use.

TYPES OF DEPOSITS

Geologists have given names to various types of glacial deposits. It will be helpful to know at least a few of them.

Glacial drift refers to all deposits that originate in glacial activity; this includes secondary accumulations, such as outwash composed of glaciofluvial debris. **Glacial till** is material deposited directly from the ice; there are several kinds. **Lodgment till** was laid down under moving ice; **melt till** was released from stagnant ice by basal melting; **flow till** was deposited by debris flowage from a glacier. Superglacial debris dumped off the edge of the ice or let down onto the ground by ablation is also classed as till.

One characteristic that distinguishes most till from other glacial deposits is poor sorting: Fragments from grains of clay to large boulders are indiscriminately jumbled together. Silt and clay form a matrix in which the larger fragments are embedded. Usually till is not strongly layered; massive heterogeneity is more the rule. Stones in till can be numerous and large, angular or rounded, and many are foreign to local bedrock terranes. Some may be glacially striated and faceted.

Scattered across many parts of upper midwestern United States and other similar glaciated regions are large surface boulders known as **erratics**. Many midwestern erratics are composed of **crystalline lithologies**, that is, rocks made up of intergrown crystals of several different silicate minerals. They are wholly different from the immediately underlying bedrock and could have come only from interior Canada, hundreds of kilometers to the north (Fig. 5.3). They are also too numerous and too big to have been imported by the Indians; the Okotoks erratic near Calgary, Canada, has an estimated weight of 118,000 tons. Noah's flood might conceivably have done the job, with the help of large icebergs, but they are generally not thought to have been part of the flood, and independent evidence of powerful flooding is usually absent. The distance is too great for transport by debris flows, which are otherwise capable. These and other bits of evidence support the conclusion that the erratics were transported by glaciers.

Some erratics have come from sources 1,000 km distant. If an erratic is so distinctive in character that its source can be unequivocally identified, it is known as an **indicator**. Some indicator sources are so localized that their stones are spread over a fan-shaped area downstream, forming a **boulder train** or **indicator fan**.

Debris flows of nonglacial origin are capable of creating deposits that resemble glacial till. Geologists may have difficulty determining the origin of such material without considering the regional setting. In Death Valley debris flows are definitely favored over glaciers by the climatic conditions, but in the nearby Sierra Nevada, which has been heavily glaciated, an unequivocal decision may not be possible. Geologists then take refuge in the vague term **diamicton**, which is a bit like saying, "Who knows?" Till-like deposits, that is, diamictons, have been found in rocks hundreds of millions to billions of years old. Where such deposits can be shown to rest upon striated, smoothed, clearly glaciated rock surfaces, they are called **tillite**, a hard, coherent, cemented till.

Rock debris initially carried by glaciers can be secondarily transported and laid down as glaciofluvial material by streams of meltwater. Such deposits display better sorting than most till (more uniform particle size) and have crude to distinct bedding, and the smoothness and rounding of their stones reflect wear during water transport. Glaciers are not good at producing rounded stones, except within the superglacial setting, where stream action can occur. Glaciofluvial gravels can usually be distinguished from stony till by comparison of the matrix material. When glaciers deposit till, they put everything down at the same time. Clay, silt, sand, and stones are jumbled together, and the matrix contains much clay and silt. Streams of water are more selective. They deposit coarser constituents first and carry clay, silt, and fine sand farther away. Consequently, the matrix of most glaciofluvial gravel is likely to be coarse sand.

Locally, glaciofluvial gravels contain spherical masses of coherent, clay-rich till, mostly less than a meter in diameter. These **till balls** form where a glaciofluvial stream has cut a steep-walled channel into till. Occasionally channel banks collapse, dumping chunks of till into the stream which treats them as though they were boulders, rolling them along its bed, rounding them, and in some circumstances armoring the outer surface with smaller stones that adhere to the sticky clay forming **armored till balls**.

Streams can carry fine suspended particles far beyond the extraglacial zone, where they may be spread over the wide, flat valley bottoms of large rivers, known as floodplains because they are inundated only at times of flood. The lower Mississippi River has extensive reaches of floodplain.

Many particles are so fine they stay in suspension all the way to the ocean, where they become seafloor sediment. Clay and the finest silt particles deposited anywhere on land within the extraglacial setting can, after drying, be picked up and transported thousands of kilometers by wind. This far-distant dust and the seafloor sediments, although supplied by glaciers, are not usually classed as glacial drift; they are too far removed from the glacial environment.

LANDSCAPE FEATURES FORMED BY GLACIAL DEPOSITION

Again, the family of forms is large, so a number of family names will be introduced. Fortunately, some will be familiar, most are simple, and some are self-explanatory.

Moraines

A **moraine** is a ridge of generally stony debris built by a glacier. It may be composed wholly of till or a mixture of till and glaciofluvial debris.

Some moraines are composed wholly of glaciofluvial deposits, possibly including an occasional glob of glaciolacustrine material. Moraines left by valley glaciers have been formed mostly by dumping and are properly classed as **dump moraines**. Although bulldozing is not a major process of moraine formation, small **push moraines** have been reported off the ends of some valley ice streams.

The planimetric configuration of a morainal ridge outlines the edge of the glacier that built it. For valley glaciers, the distal part of the morainal ridge is usually convex down-valley, and the up-valley extensions are parallel to the valley walls. The convex part is an **end moraine**, and the up-valley extensions are two **lateral moraines** (Fig. 8.1). End and lateral moraines grade into each other.

An end moraine lies largely on the valley floor and extends only a short distance up the valley sides. Lateral moraines lie on the valley walls. The extent of an end moraine is limited by the width of its valley, but a pair of lateral moraines may continue up-valley for kilometers. Where ice streams have emerged onto a gentle, featureless slope at the foot of the mountains, lateral moraines can form large, free-standing ridges that effectively extend the valley beyond the mountain front.

If an end moraine lies at the point of farthest advance, it is a **terminal moraine**. Recession from this terminal position has usually been interrupted by pauses and short readvances. At each stabilized position, another end moraine may have been built. These are **recessional moraines** (Fig. 8.2), and a valley may have fifteen or twenty. The size of each reflects the duration of ice stability. If it was long, the recessional moraine may be nearly as large as the terminal moraine; sometimes it is even larger. Larger recessional moraines are usually accompanied by recognizable lateral moraines on the valley walls. An airborne view up a valley with a complex of nested end and lateral moraines can be impressive (Figs. 8.3 and 8.4).

Lateral moraines built by glaciers that have emerged from mountain valleys can be huge. Some of the tallest lateral moraines attain heights of 200 to 300 meters (Fig. 8.4). They are usually composite, having been constructed during more than one glacial advance. Although large moraines are the product of long-enduring stabilized ice positions, the rate at which ice was moving up to that position and the amount of debris being carried also affect morainal bulk. Clean ice simply doesn't build a moraine, except perhaps a little one by bulldozing, no matter how long it lingers.

Moraines formed by continental ice sheets have simpler configuration and gentler relief. In close view they may appear almost straight, but their larger form is primarily lobate. The front of the continental ice

Figure 8.1. A side-looking aerial photo of a double-crested lateral moraine (A and B) built by two separate advances of Ladd Glacier on the north slope of Mount Hood, Oregon. The sharpness and freshness of the inner morainal ridge (B) suggests an age within historical times. The somewhat dirty ice-tongue terminus (I) lies in the upper left-hand corner. Morainal ridges of this type are built primarily by the dumping of superglacial and englacial debris off the lateral margins of the glacier (see Plate VIII). (U.S. Geological Survey photo by Austin Post.)

sheet was not a simple uniform line; rather, it was a changing assemblage of tonguelike projections, with tongues of different sizes moving forward at different rates and times. During a late phase of the last Ice Age, east-central Illinois was invaded by a mass of ice that moved southward down the Lake Michigan trough and spread over the east-central part of the

Figure 8.2. Remnants of three (1, 2, 3) arcuate recessional moraines and a recessional moraine complex (MC) are shown on this aerial photo of the terminal area of Iliamna Glacier, Alaska. Glaciofluvial outwash material fills the swales between the moraines, which are cut through by outwash channels. Most recessional moraines do not display such perfect symmetry and regular configuration. The ice edge was remarkable in its evenness and uniform curvature when they were formed, probably within the past century or two. The area outside the outermost moraine is a beautiful outwash plain, characterized by many braided stream channels. (Photo by Bradford Washburn.)

state, maintaining an expanded lobate form and leaving end moraines of that configuration (Fig. 8.5).

Although continental sheets deposit terminal and recessional moraines, they do not form lateral moraines. If there is an analogous feature, it might be better termed a **marginal moraine**, as distinct from moraines deposited at the apex of a tongue. Moraines deposited between two adjacent, concurrently active ice lobules are described as **interlobate moraines**.

Figure 8.3. One of the most impressive sets of moraines built by a valley glacier in North America. You are looking southeast up the valley of Green Creek on the east side of the Sierra Nevada of California, in the Bridgeport Basin area. Lateral moraines of at least three phases of glaciation (1, 2, 3) and one end moraine (EM) of the youngest phase (3) are clearly evident. We see here how a glacier has literally extended its canyon outward beyond the front of the mountains by building confining ridges of glacial debris rising nearly 300 meters above the valley floor. (Photo by John S. Shelton.)

Continental moraines can be large. The Bloomington Moraine of central Illinois is 25 to 30 km wide and more than 300 km long. It is only 20 to 60 meters high, however, and the eye must be trained to recognize such broad, low features, even in a prairie landscape. The subdued relief is attributable partly to the high clay content of the Bloomington till, which promotes secondary **mass creep** and flow that has smoothed and gentled the surface; deposition of much of the moraine under the thin edge of the ice sheet by lodgment was also probably a

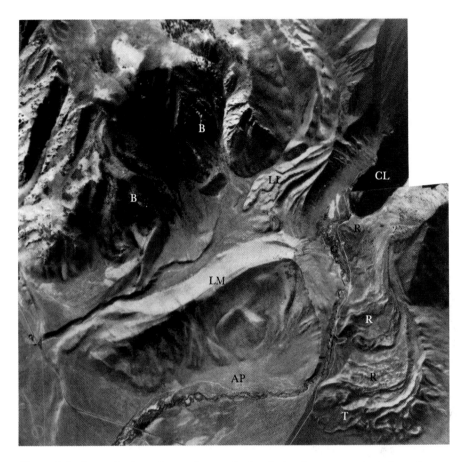

Figure 8.4. A morainal complex at the mouth of a glaciated canyon, such as shown here on vertical photos of the Convict Creek area on the east side of the Sierra Nevada, California, can tell an interesting story. Deposits of two principal phases of glaciation are involved. In the earlier phase an ice tongue from Convict Canyon (upper right) above Convict Lake (CL) moved in a straight path along the base of the bedrock (B) mountain mass (upper left) and built a huge lateral moraine (LM) along its outer margin, which rises fully 300 meters above the piedmont alluvial plain (AP). As this ice tongue receded, a marginal meltwater stream cut a gap (G) in the lateral moraine, which was then occupied, deepened, and widened by Convict Creek after ice retreated up the canyon. When the second-phase advance occurred, the ice tongue followed this new course and deposited the succession of terminal (T) and recessional (R) moraines (lower right) outside the older high lateral moraine. At its maximum thickness ice of this last advance shoved a thumb down the route of the earlier advance, forming a complex of lobate laterals (LL). A bulky recessional moraine of the last advance dams Convict Lake. The sharpness and irregularity of the younger moraines contrasts nicely with the smoother, subdued topography of the earlier moraines, an indication of significant age difference. (Vertical aerial photos, courtesy of Geotronics, October 19, 1944.)

factor. Other end moraines built by continental sheets by dumping and richer in sand and stones display more marked local relief and greater topographic irregularity. Such moraines are more easily identified, the

Figure 8.5. This map showing moraines (black) of the last episode of glaciation in part of midwestern United States demonstrates the lobate configuration of the edge of the continental ice sheet. At least four lobes, as labeled, were involved in creating this pattern. (After J.C. Frye and H.B. Willman, 1973, "Wisconsinan climatic history interpreted from Lake Michigan lobe deposits and soils." Geological Society of America Memoir, 136, 135–52.)

Kettle Moraine of east-central Wisconsin being a good example. It has a distinctive **knob-and-kettle topography** that gives the moraine its name.

GROUND MORAINE.   Glaciers deposit debris as they advance or recede, as well as when the ice margin is stabilized. These deposits of **ground moraine** compose **till sheets** usually only a few meters thick that mantle extensive areas. The till can consist of lodgment debris, or ablation material, or a combination of the second topping the first.

Figure 8.6. Understanding of older glacial features can be aided by investigation of forms developed by existing glaciers. This recently deglaciated area just off the edge of Malaspina Glacier, Alaska, displays a strong linear ridge-and-groove pattern on ground moraine. Large areas of ground moraine with this type of marking have been left by the continental ice sheet in Montana and the Dakotas of the United States. Most ridges are molded rather than erosional, consisting of till pressed into grooves in the bottom of the ice sheet. (August 15, 1951.)

The surface of most ground moraine is a disorganized assemblage of low, rounded swells and shallow swales. In some localities a more organized and distinctive pattern of hollows (**flutes**) and ridges is present (Fig. 8.6). Although this has been called **fluted moraine**, the term **ridged moraine** is preferable. The hollows exist because of the ridges, usually about a meter high, made of till that was molded into grooves carved into the base of the ice by rock outcrops or boulders anchored on the bed some distance upstream. These **groove casts** extend perpendicular to the glacier's margin; within groups the arrangement is parallel to slightly radial.

Some areas of ground moraine display much larger groove casts, up to 20 km long and 100 meters high, which separate **megaflutes**. Groups of megaflutes have also been formed where a glacier moved across exposures of relatively soft bedrock. Shorter, narrow, and more sharply

Figure 8.7. Part of the dirty, crevassed terminal area of Schwan Glacier (SG) in southeastern Alaska appears in the upper right corner of this near-vertical aerial photo. The short, narrow linear ridges on the recently exposed subglacial floor to the left, beyond the ice edge, are crevasse casts (CC) or fillings, some showing the same orientation as crevasses within the existing ice. At upper left is a nice valley train (VT) of glaciofluvial debris with braided channels. The winding, narrow ridge crossing the lake (L), middle left, looks like an esker but may be a morainal deposit. (Photo by Bradford Washburn.)

crested ridges of till on ground moraine, without the consistent orientation of groove casts, may be **crevasse casts**, formed of till that was squeezed into crevasses that penetrated to the floor of thin glacier ice (Fig. 8.7).

A few till sheets in North America and Europe display an unusual feature. Near the margin is a succession of 1- to 5-meter-high ridges separated by shallow swales 30 to 200 meters wide. These forms extend for several kilometers essentially parallel to the outer edge of the sheet. The swells resemble broad subdued waves, and, where numerous, they constitute **washboard moraine**. Distance between swells in any group can be fairly consistent, supporting the possibility that each swell–swale couplet is of annual origin. The swell presumably reflects deposition under the more nearly stabilized ice margin of winter, and the swale represents the recession that occurred during summer melting. Measurements made on such features over an area in Iowa suggest an ice-edge recession of about 1.6 km in fifteen years, or a little over 100 meters per year. Ground moraine is more characteristic of ice sheets than of ice streams. Much fertile farmland in midwestern United States and northern Europe lies on areas of ground moraine.

Drumlins

A striking product of glacial deposition in some areas deeply mantled by ground moraine is groups of elongate, streamlined hillocks called **drumlins**. These are not to be confused with the erosional rock drumlins earlier described. The upstream end is blunt and steeper than the tail, like a teardrop. The sides can be even steeper than the nose. Some depositional drumlins have forked tails or grooved sides, and large ones may have smaller drumlins perched on their flanks. The long axis is oriented parallel with the direction of ice movement. Depositional drumlins are mostly 10 to 20 meters high, tens to 500 meters wide, and from one-half to several kilometers long.

Many are composed of compact lodgment till or have a veneer of such till over a core of other material, such as older till, glaciofluvial debris, or even bedrock. Such forms are **cored drumlins**. Drumlins have clearly been shaped by overriding ice, but how much of the streamlining is due to deposition and how much to erosion is still a matter of debate.

Drumlins are found in groups or swarms (Fig. 8.8) containing as many as 10,000 individuals. Slight outward radial dispersal of drumlin axes is apparent in some assemblages. Drumlin fields exist in Ireland, England, the Canadian Rockies and prairies, Wisconsin, Michigan, upstate New York, western Washington, and New England. Some islands in Boston Harbor are drumlins, and that citadel of United States history, Bunker Hill, is a drumlin. They are a product of ice sheets; rarely do mountain ice streams build them, except where they have spread out on gentle piedmont plains, as in Switzerland.

It is speculated that drumlins form near the edge of an ice sheet under conditions of unusually heavy debris load in the basal ice. The ice may be greatly slowed, but it must be moving enough to shape the deposit as it is laid down. Lodgment till, rich in fine particles and containing considerable water, expands under stress, becoming more mobile. This

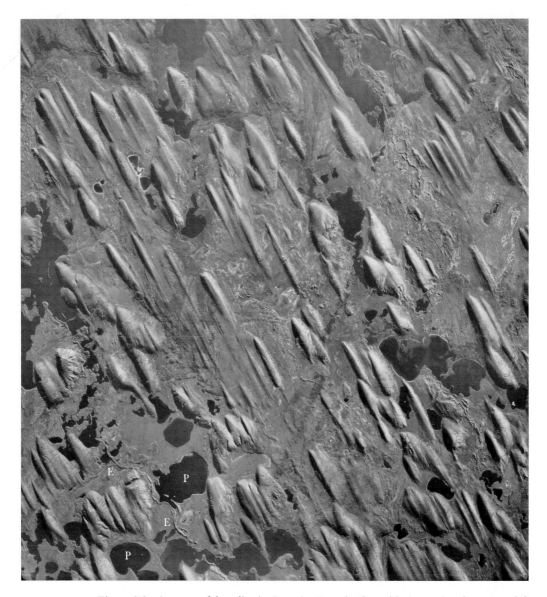

Figure 8.8. A swarm of drumlins in Ontario, Canada, formed by ice moving from upper left to lower right. As drumlins go, many of these are unusually long and narrow, but the classical blunt upstream nose and tapering narrow tail are apparent on most. There can be little doubt that ice moving toward the lower right shaped these hillocks. Several eskers are also shown, the longest and best in the lower left (E). Dark spots are ponds (P). (Photo [A14509 5] reproduced from the collection of the National Air Photo Library by permission of Energy, Mines and Resources, Canada, Her Majesty the Queen in Right of Canada.)

behavior is thought to facilitate development of the streamlined form. Sections cut through drumlins by roads, railways, or other excavations may show a lamination in the till that conforms with the external shape, suggesting that those drumlins were built in accretionary fashion, layer upon layer. An airborne view of a drumlin field obliquely illuminated by a low sun demonstrates the remarkable consistency in size, symmetry, and orientation of the forms (Fig. 8.8).

Ice-contact features

When contact with an ice mass has controlled or strongly influenced the shape, configuration, and character of any glacial depositional feature, it is classed as an **ice-contact feature**. Many ice-contact features are of glaciofluvial origin, but not all. An exception is the inner flank of a dump moraine, which is an ice-contact face secondarily modified by slumping, as are most ice contact faces. Contrariwise, some glaciofluvial forms, outwash plains and valley trains, for example, are not ice-contact features.

One of the most ubiquitous ice-contact forms is a **kettle hole**, or simply **kettle** (Fig. 8.9). It is a closed topographic depression, usually

Figure 8.9. Kettle holes are an abundant ice-contact feature in many youthful glacial deposits. This photo shows a typical example formed within a moraine in Minnesota by the melting of a good-sized block of ice. The gentle bank, near-circular outline, and smooth shoreline configuration suggest that the ice block was completely buried.

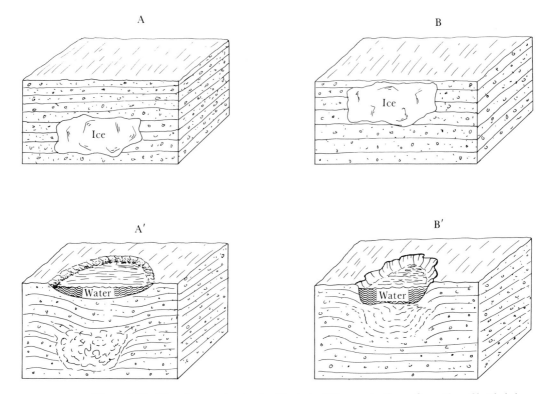

Figure 8.10. These interpretive sketches illustrate differences in the configuration of kettle holes formed by deeply (A, A') and shallowly (B, B') buried ice masses. Kettle holes with steeper banks and more irregular outline generally indicate shallow or incomplete burial of the ice.

simple in configuration and partly filled with water, that forms by melting of a buried block of glacier ice. Kettles are common in glaciofluvial deposits but also occur within accumulations of till. The size, shape, and depth of the ice block controls many aspects of a kettle hole. The more deeply buried the block, the weaker its influence (Fig. 8.10). Kettle holes of smooth outline and gently sloping sides formed over ice blocks more deeply buried than those that created kettle holes with irregular outlines and steep sides.

The last type of kettle hole may have little ramparts of jumbled, coarse, bouldery material in spots along its margin. Ice blocks forming such kettles were not completely buried; their upper parts projected above ground to a significant height. Ice entombed in moraines or outwash deposits comes from the terminal part of a glacier and is usually rich in rock debris. Most of this material is deposited on the kettle hole floor as the ice melts. In the instance of incompletely buried ice blocks, however, some debris falls onto the ground around the edge of the ice (Fig. 8.11), making little bouldery ramparts.

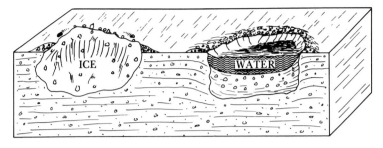

Figure 8.11. The bouldery rampart (right) around the edge of some kettle holes is formed by the shedding of debris from a partly buried block of dirty ice (left).

Kettle holes may be only a few meters across or many kilometers in maximum dimension. Depths are from 1 or 2 meters up to 50 meters. Crudely circular shapes predominate, but they can be highly irregular. Some kettles are so elongate as to merit the term **kettle valley**. Groups of kettles can form linear or gently sinuous strings. Some kettles in outwash deposits have formed over individual blocks of ice carried from the glacier by streams (Fig. 8.12). Glaciofluvial deposits can also be spread across extensive areas of stagnant ice which, upon melting, produces a richly kettled surface called **pitted outwash**.

Kettle holes add to the scenic attractiveness of any landscape. Few people realize that the pond or lake near their home or summer cabin that they enjoy so much may have been created by melting of a buried block of glacier ice.

Let us now consider knobs instead of depressions. Imagine a crudely cylindrical hole of good size that penetrates to the bed of sluggish ice composing the thin marginal part of a glacier. Let a superglacial stream carrying a load of sand and gravel pour into that hole. In time the hole will be filled with glaciofluvial material. When the ice melts, an isolated, freestanding conical hillock, with slopes at the angle of repose for un-consolidated rock detritus (about 30°), will be left. This is a **kame**. Hillocks of similar origin but more irregular form are also kames, as are ridges or mounds of glaciofluvial material deposited into any mold defined by walls of glacier ice.

In Yellowstone National Park an unusual type of kame was formed when warm vapors and water from hot spots in the volcanic rocks of that geothermal area melted holes through the thin ice of the receding Yellowstone cap. Such holes were typically 200 or 300 meters in diameter and 50 to 70 meters deep. Meltwater streams filled the holes with sand and gravel. All the while, the underlying hot spot was emitting vapors and hot waters rich in mineral matter, especially silica (SiO_2), which

Figure 8.12. A small kettle hole in outwash debris near the edge of Malaspina Glacier, Alaska, formed the day before yesterday, geologically speaking, by melting of a small completely buried block of glacier ice.

accumulated in open spaces between fragments of the sand and gravel and firmly cemented the deposit. Later the ice melted, leaving freestanding, steep-sided pillars of hard rock known as **geothermal kames**.

Picture a large stream of meltwater running laterally along the edge of a valley glacier. Assume that the stream is powerful enough to pick up and redistribute all the debris being dumped off the ice that would otherwise be forming a lateral moraine. The stream spreads this material over a wide, flat-floored channel, creating a glaciofluvial train many meters thick. On one side is glacier ice, and on the other is the valley wall. When the glacier shrinks, the train is left as a **tread** bounded by a steep **riser** on the glacier side where material was deposited against ice. The form is that of a terrace; specifically, it is a **kame terrace**. Kame terraces form best along the lower reaches of deteriorating ice streams or along the margins of masses of stagnant ice residually trapped in bedrock valleys.

The ice-contact face of a kame terrace may display projecting, peninsula-like ridges. These are puzzling until one recalls that the face was built against a glacier. The projections probably represent fillings of open marginal crevasses in the ice against which the kame was deposited.

Another unusual ice-contact feature is an **esker**, a long, narrow, steep-sided ridge of glaciofluvial sand and gravel inhabiting a glaciated area. Eskers tend to follow valleys and lowlands, carefully picking a course between obstacles. Yet some eskers display the surprising ability to go uphill across transverse ridges, usually crossing through a low saddle or gap. Eskers ascend and descend as much as 250 meters. Most eskers are a single ridge, but some are braided or branched like river tributaries. Although their courses may be irregular, even serpentine (Fig. 8.13), the trend of long eskers is parallel to the direction of ice movement. The path of an esker can be locally interrupted by short erosional gorges. Longer channels, presumably eroded by subglacial streams, are known as **tunnel valleys**.

Most eskers are a few to 20 or 30 meters high, but some are huge, attaining heights exceeding 200 meters. They may be anything from a few kilometers to 500 km long. Some terminate at their distal end in a glaciofluvial fan built onto an outwash plain (Fig. 8.14) or in a delta built into an extraglacial lake.

Internally, they are composed largely of well-sorted glaciofluvial sand and gravel with waterworn stones. Some locally sport a thin mantle of coarse, angular, jumbled ablation debris. The steep sides of an esker ridge are near the angle of repose, about 25 to 30 degrees. The upper meter or two of the mantle on these slopes usually looks like slumped debris.

Some eskers are beaded, meaning they pinch and swell in a regular pattern. The beads may possibly be of annual origin, reflecting greater summertime deposition and widening of an ice channel indenting the front of a receding glacier. The various reaches of long eskers were probably not all formed at the same time. Many eskers have grown headward in successive increments as the ice mass shrank, so that the outer end is older than the head.

Eskers are thought to form during the waning phase of a glacial episode, mostly in association with ice sheets. Thin, sluggish, deteriorating ice appears to be the condition most favorable for their development. Much meltwater and abundant debris are desirable requisites. The ice cannot have moved much after the esker formed; otherwise it would have been destroyed.

Eskers consist of deposits laid down by subglacial streams flowing in ice-walled channels. Some channels may have been open to the sky. Others were clearly closed by ceilings because they functioned like pipes under pressure in a water system. Only within completely enclosed passages could water go up hill and down dale, as some eskers do. Because

Figure 8.13. A side-looking aerial view of a late Pleistocene esker (E) sinuously winding across a ground-moraine plain (P) in central Minnesota. A left-to-right country road (R) crosses the near end of the esker. Above the road and to the right of the esker is a hayfield (HF); the dots are piles of recently mown hay. The crest of the esker is bare, but the flanks bear trees (T). An esker of this type is presumably formed by glaciofluvial deposition in a subglacial tunnel or an open ice-walled gorge within a receding, nearly inactive ice sheet. (From W.S. Cooper, 1935, The History of the Upper Mississippi River in Late Wisconsinan and Postglacial Time, Minnesota Geological Survey Bulletin 26, 116 pp.)

of glacier movement and solid-state flow, the walls of ice channels or tunnels pressed in on the subglacial stream. It had to be powerful enough to cut back the ice as fast as the walls moved inward. Although a large amount of sand and gravel was laid down within the tunnels, as demonstrated by the size of many eskers, complete choking probably did not generally occur. As deposits filled the tunnel, the stream made room for itself by cutting back walls and ceiling.

In rural areas where many backcountry roads are surfaced with gravel,

Figure 8.14. In this side-looking aerial photo the rounded smooth slope in the upper right (I) is the snout of Woodworth Glacier, Alaska. The narrow, irregular ridge (E) extending outward to a glaciofluvial fan (F) is an esker of recent origin. Its outer part at least was formed at a time when the edge of the ice was up against the glaciofluvial fan, for the sharp, steep slope at the fan's inner edge is an ice-contact face (IC). The esker becomes braided (B) about one-third of the way toward the ice, suggesting that there it was formed near the receding edge of the ice by streams flowing in interconnected ice-walled channels and not in a single ice tunnel. About halfway to the current ice front, the stream was in an erosional mode, and from there back it cut channels (C) into the subglacial floor. The ground-moraine sheet (GM) to the right of the esker is nicely ridged, and at least two of the ridges can be seen to start just to the lee of large boulders (B) near the ice front, just left of the rightmost proglacial lake (L). This splendid picture is a veritable textbook of glacial features. (Photo by Bradford Washburn.)

eskers are prized as a source of road material. Nearly any reasonably accessible esker will harbor more than one gravel pit. Road cuts and gravel pits provide good exposures for study of their internal makeup and structure. Bush pilots flying across the central Canadian wilderness have used long eskers of consistent trend as guides for ground-contact navigation. Viewed from the air, they stand out vividly amid other features of the glaciated terrain and look like large railroad embankments.

EXTRAGLACIAL FEATURES

As a glacier recedes, the discharge of meltwater by glaciofluvial streams may continue to be high, although the supply of debris may decrease. Streams are tireless eroders and readily dissect the outwash plains they have previously constructed, incising channels into them.

Glacier recession does not usually occur at a smooth, steady pace. It is an uneven affair, and this unevenness is reflected in the behavior of glaciofluvial streams. Dissection of outwash deposits is likely to proceed in an irregular, interrupted fashion. Periods of incision are separated by intervals when stream channels are widened rather than deepened, or even partly refilled with new glaciofluvial deposits in response to a small readvance of the glacier. When incision is renewed, part of the wide, flat channel floor is left as a step, or **terrace**, alongside the new channel. A succession, or flight, of terraces at several different levels along channels cut into outwash plains is normal. Valley trains can also display flights of **outwash terraces**.

Glaciolacustrine features

Once the basin of a proglacial lake is breached by a through-flowing stream, the beds of lacustrine sediment are dissected and exposed, possibly revealing varves. Varved sections many tens of meters thick record geological time intervals of hundreds or even thousands of years. The measurement and plotting of the thicknesses of individual varves within sequences can look like a stock market graph (Fig. 8.15). Occasionally, conditions along the front of a receding glacier allowed a succession of extraglacial lakes to overlap each other in time. Within such successions the youngest varves deposited in an older lake have been correlated with the oldest varves laid down in younger lakes. Correlations are aided by the occasional varve that is so distinctive in character – thickness, color, texture, or special constituents like volcanic **ash** or **diatoms** – that it is unique in any sequence.

Through painstaking study, measurements, and correlations, composite varve plots covering thousands of years have been constructed. The longest sequence recorded is in Scandinavia and embraces almost

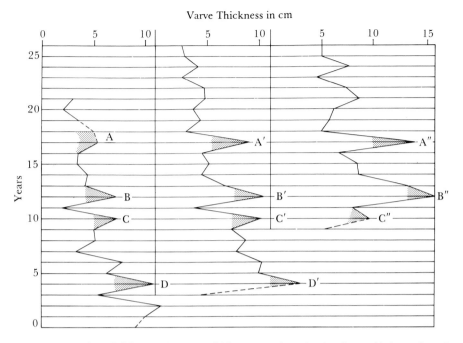

Varve Thickness in cm

Figure 8.15. Plots of differences in varve thicknesses at three sites in the Stockholm region of Sweden, showing how varve-sequence graphs are correlated. Varves of unusual thickness are emphasized by lettering and darkening. (After G. De Geer, 1912, "A geochronology of the last 12,000 years." Compte rendu de la XI:e session Congrès géologique international, Stockholm 1, 241–53.)

17,000 years. Not everyone regards such lengthy sequences as valid, although 12,000 years of the 17,000-year sequence have been confirmed by **radiocarbon** (^{14}C) dating. Many investigators readily accept more local and shorter composite varve sequences and use them to date archeological remains, to measure the rate of geological processes like sediment accumulation or the progression of weathering and erosion, and to date specific geological events. Before the advent of **carbon-14** dating, varves were one of the best means of obtaining time measurements over the past few thousands of years.

Varves are still being used in productive ways. For example, a recent study of a 7,000-year varve sequence cored from the bottom of a Minnesota lake records the history of changes in **declination** of Earth's magnetic field. This is possible because some mineral grains in the varved sediments are magnetic and acted like tiny compasses as they accumulated on the lake floor. Their axes point toward the location of the north magnetic pole at the time they reached the bottom. This varve sequence shows how the north magnetic pole has wandered during the past 7,000 years.

Glacioeolian features

A person naturally associates dunes with shorelines of an ocean or large lake, where wind blowing from the water picks up the sand on beaches and carries it inland. Deserts are also favored sites of dune formation because of strong winds, sparse vegetation, and liberal supplies of sand provided by alluvial deposits repeatedly refreshed by cloudburst floods. An active glacial outwash plain is like a desert in its bareness, strong winds, and frequently renewed supply of raw glaciofluvial debris, which make it a fertile source of **glacioeolian** material (glacial debris reworked by wind.)

It is not surprising, therefore, that many old outwash plains, left by ice-age continental glaciers in Europe and North America, are bordered by large sand-dune sheets created when the plains were active. In the intervening thousands of years the dunes have become stabilized and covered by vegetation, and their topography and relief have been softened by erosion. It may take an experienced eye to recognize that an area of gently rolling, bumpy terrain was once a sheet of active, migrating dunes. In places where vegetation has been destroyed recent winds may have excavated hollows, known as **blowouts**, and a little present-day sand movement may be occurring in them.

In blowouts, one can see how well-rounded and uniform in size and shape are the grains composing wind-blown sands. Road cuts may also expose the characteristic bedding of dune deposits, inclined up to 30° in various directions. The sands are homogeneous and usually uniformly oxidized to a rich brown color because of their high porosity. Vegetation growing on old dunes differs from that on surrounding nondune areas. Farmers know that only certain crops do well on dune sands because they are so permeable.

Modern housing tracts are laid out on old dune sheets because the rolling terrain makes an attractive setting and the underlying sands are easy to grade and excavate. Residents would probably be surprised, even shocked, to learn they were living in the midst of an assemblage of old sand dunes, but the signs are there. Lawns have to be watered with unusual frequency because the underlying sands drain so readily, and children and pets track sand instead of mud into the house.

Close inspection of boulders on an outwash plain from which dune sands were derived – even an old stabilized plain – would almost surely reveal that the glacierward sides of some display small flutes, grooves, pits, etched structures, possibly polish, and smooth or gently curving facets. These are the characteristics of wind-blasted stones or ventifacts (Fig.

7.9). The strong winds, sparse vegetation, and abundant saltating sand of an outwash plain are ideal for creating them. Ancient ventifacts formed during the Ice Age are scattered over the glaciated areas of both North America and Europe. They are not hard to find in the right places, once our eyes learn to recognize the characteristic features.

Dust that goes into suspension over active outwash plains can travel hundreds of kilometers before coming to rest. There it may simply be dispersed and lost among other surface materials, or it may settle into the ocean as part of the seafloor ooze. Where a large amount accumulates on land, it can form a sheet many tens of meters thick. Such deposits are loess. In parts of interior China, loess, believed to have come from a desert source rather than from glaciers, is nearly 300 meters thick.

In the prairielands of central United States, a lot of the finest glacial flour remained suspended in rivers flowing south from the ice sheet into wide valleys, and much of it ended up on their **floodplains**. After drying, it was picked up by strong prevailing westerly winds, transported, and laid down as a blanket of loess over lands immediately to the east. This blanket is coarser and thicker adjacent to the river, especially those parts having the widest floodplains (Fig. 8.16). Locally, loess forms riverbank bluffs; Council Bluffs, on the Missouri River in Iowa, is a historic example.

Loess deposits are fine-grained, well-sorted, homogeneous, and massive, mostly without bedding. Being porous, they oxidize quickly to a uniform tan color. Near the surface, secondary patterns of staining by weathering and disruption by plant roots and animal burrows perturb this homogeneity. Shells of fossil land snails embedded in loess have been studied for evidence of the climatic environment attending its deposition. Where some of the loess particles are composed of limestone, secondary solution and redeposition of calcium carbonate ($CaCO_3$) have created little nodules (**concretions**). Some have shapes that suggest small human figures or body parts; these are the famed **loess children**.

Insects, rodents, and humans find loess easy to excavate, so they dig chambers, tunnels, and caves into the face of loess bluffs. Although it excavates easily, much loess is coherent enough to support free-standing walls and ceilings. Peasants of inland China have lived for generations in loess caverns. Mao Zedong once occupied one as his headquarters. They provide inexpensive, well-insulated, and weatherproof housing, but with one great disadvantage. Inland China has large earthquakes. On occasions, they shake down the bluffs, collapse the caverns, and take a huge toll of human life. It's a cold, windy region, so peasants are willing to risk death by earthquakes just to be warm and dry.

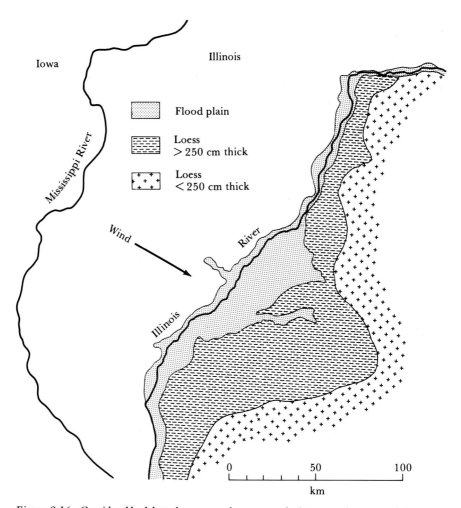

Figure 8.16. Considerable debate has occurred among geologists as to the source of dust composing deposits of loess in midwestern United States. Relationships depicted on this generalized map indicate that in Illinois, at least, the windblown silt for extensive loess deposits has come from adjacent river floodplains. The reasoning is based on the fact that the most extensive deposit of thick loess (greater than 250 cm) forms an east-southeast-projecting lobe just east-southeast of the widest part of the Illinois River floodplain (stippled). Strong prevailing winds, during glacial times, from the west-northwest picked up glacial silt from the floodplain and deposited it over the land to the east-southeast.

Farmers know that weathered loess makes excellent agricultural soil; it is uniform in texture, easy to cultivate, and rich in minerals, and it stores moisture well. Some of the most productive soils in North America, and in a belt extending eastward across central Europe into Asia, are underlain by loess. The lava flows of eastern Washington State would be agriculturally barren were it not for a capping of loess, which makes the area one of North America's richest wheatlands.

Miscellaneous products of glacial deposition

Under this heading we will discuss a small number of depositional features that do not fit well into any preceding category.

MINOR MORAINAL MARKINGS. The bottom of a glacier near its margin may have been grooved to a depth of a meter or two by projecting bedrock outcrops or firmly seated boulders on its bed. Lodgment till deposited under the ice is easily squeezed into such grooves. When the ice melts, linear ridges of till are left, usually not much more than a meter high and 2 or 3 meters wide at the base. These ridges extend leeward from the engraving obstruction in the direction of glacier movement, gradually becoming lower and eventually disappearing. The decreasing height reflects shallowing of the groove caused by bottom melting. The best places to look for such features are just beyond the snouts of currently receding glaciers (Figs. 8.6 and 8.14). It is usually possible to trace a till ridge back to the boulder or outcrop that carved the ice groove (Fig. 8.17). Most longitudinal linear markings on ground moraine deposited by ice sheets are probably of this origin.

MORAINE-DAMMED LAKES. Lakes occupying glacially excavated bedrock basins have already been discussed. Many such basins have rims elevated by morainal deposits that measurably increase their capacity. Of the 433 meters depth of Lake Chelan in Washington State, 115 meters are attributed to morainal damming and 318 meters to bedrock excavation. Many cirque and valley-floor lakes occupying bedrock basins have supplemental morainal dams. Some glacial lakes have only morainal dams, but they are hard to identify without extensive sounding and drilling, which are expensive and seldom employed.

Among the more picturesque of predominantly moraine dammed water bodies are **piedmont lakes**, lying at the foot of lofty glaciated mountain ranges. Jenny Lake in Jackson Hole, Wyoming, below the Tetons, is an example well known to tourists. The attractiveness of such lakes is enhanced by the backdrop of high, rugged, usually snowcapped mountain peaks.

Piedmont lakes are formed where an ice stream descending a steep mountain valley arrives at the foot of the mountains and spreads out over the adjoining smooth, gently sloping piedmont surface, forming a semicircular bulb of ice (Fig. 2.2). Eventually, the ice stabilizes and starts to build a moraine around its edge by the dumping process. When it recedes, the abandoned moraine encloses a basin lying tight against the mountain foot, in which a lake forms.

Figure 8.17. This view downstream from near the terminus of recently receded Saskatchewan Glacier in Jasper National Park, Canadian Rockies, shows a foreground boulder (B) so firmly anchored that it was not moved by the thin overriding ice. Instead, the boulder carved a groove into the bottom of the glacier into which wet, mobile till was molded to form the slightly curved ridge (R) extending about 30 meters to the lee of the boulder.

Moraine-dammed lakes can be formed wherever moraine deposition creates a closed basin and water supply is adequate. Most of them are not long-lived because outlet streams cut rapidly through the dam, and incoming streams carry so much debris that the basins are rapidly filled.

Most existing moraine-dammed lakes are a product of the most recent episode of glaciation and are therefore at most 15,000 to 20,000 years old.

ICE-DAMMED LAKES AND GIGANTIC FLOODS. Human history and geological evidence record instances of lakes, some very large, impounded behind dams of glacial ice. This happens mostly where an advancing glacier moves across or partway up a river valley, blocking the drainage. Smaller lakes can be impounded within vales or tributary canyons along the lateral margins of large ice streams in trunk valleys. Much of the water may be supplied by melting of the glacier itself or of other nearby glaciers. Some marginal water bodies of modest size, such as Tulsequah Lake, dammed by the glacier of that name in the Coast Range of British Columbia, Canada, discharge catastrophically each summer. The Himalaya Mountains of Asia are notorious for the devastation created by floods arising from collapse of ice dams impounding lakes in upper reaches of some major rivers.

Neither the U.S. Army Corps of Engineers nor the Bureau of Reclamation uses ice to build dams; it is not a good structural material. Once a lake fills and an outlet stream overtops an ice dam, erosion occurs so rapidly that destruction of the dam and flooding are almost assured. Similar trouble arises if the lake gets so deep that the dam floats loose from its footings or the water pressure gets high enough to cause the ice to yield by solid flowage. A floated dam can collapse quickly, generating a catastrophic flood possibly laden with huge icebergs. Geological evidence testifies that huge floods of this type have occurred repeatedly because of ice dams formed by glaciers of the most recent Ice Age, 13,000 to 20,000 years ago. Similar dams and floods were also undoubtedly created by earlier glaciations, but the evidence is obscured.

One of the greatest of these catastrophes was the Spokane Flood, which originated near Pend Oreille Lake, Idaho, and swept 880 km across the lava plains of eastern and central Washington State and down the Columbia River to the Pacific Ocean. The floodwaters came from glacial Lake Missoula, which covered 7,700 km^2 of Idaho and western Montana with 2,500 km^3 of water to a greatest depth of 600 meters. This lake formed when the Purcell ice lobe advancing south from Canada stuck a thumb up the Clark Fork valley in northern Idaho, blocking it with an ice dam at least 600 meters high. Eventually, waters of Lake Missoula either overflowed the dam, caused it to yield by solid flowage, or floated it off its footing, unleashing the huge torrent of water that became the Spokane Flood.

The initial models of this event postulated total collapse of the dam creating a single huge flood. It was also recognized that repeated episodes of advance and recession by the Purcell ice lobe could have caused several floods during the past few hundred thousand years. The Spokane Flood is only the latest of these occurrences, and it is now proposed that even it consisted of a large number of separate flood events.

From estimated rates of discharge it is calculated that Lake Missoula emptied completely in just about two days. Owing to temporary impounding of water behind narrow gaps along the flood's path, flooding probably continued through the devastated area for two to four weeks. Velocities attained 70 km per hour in local confined channels, such as the Clark Fork canyon. The discharge, expressed in cubic meters per second, was about ten times the present combined discharge of all the great rivers of the world.

In the early 1980s a different concept of dam behavior and flood sequences evolved. It was proposed that the impounded Lake Missoula waters buoyed up the ice dam and broke the seal at its base, permitting underflow through rapidly enlarging subglacial channels to create a huge, catastrophic flood. Thus, the dam was not destroyed, and it is pictured as settling back into place after most of the water had been discharged. This would be like loosening the cork of an inverted water bottle and then jamming it back into place before the bottle is completely emptied. It is calculated from present-day discharges of drainage areas tributary to Lake Missoula that the lake could have refilled to its upper level in about eighty years, give or take perhaps twenty years, making reasonable allowances for increased glacier-age runoff. The time required in each instance would depend upon how completely the basin was emptied by the preceding discharge. Once the lake was refilled, the whole sequence could be repeated.

The dam was a broad lobe of ice about 65 km wide, extending well south and west of the Clark Fork valley. Such a mass would have considerable integrity and, by means of internal solid flowage, could heal itself and close down subglacial channels as the flood discharge diminished. Initial opening of the channels may have been facilitated by solid ice flow under the hydrostatic head of the lake waters.

Rhythmically layered flood deposits (Fig. 8.18), formed in backwater enclaves along the flood route in northeastern and south-central Washington, are interbedded with Mount St. Helens ash, loess, and varved lacustrine layers. From this sequence it is concluded that at least forty to sixty episodes of catastrophic flooding occurred during the Spokane event, not just one. The known age of the ash layer and radiocarbon dates on the interbedded materials show that Lake Missoula existed for

Figure 8.18. Rhythmically layered sand silt beds laid down as backwater deposits by catastrophic Spokane-type glacial floods as exposed in Burlingame Canyon of the Walla Walla Valley in south central Washington State. The ash layer (A) came from an eruption of Mount St. Helens 13,000 years ago. This date is consistent with others derived from similar deposits. These and corresponding sequences of beds in other localities suggest that floods came repeatedly from Lake Missoula between 15,300 and 12,700 years ago, because its ice dam became alternately unsealed and resealed, like a cork in a bottle. The little 2-meter cliff at the top is loess. (Photo by R.B. Waitt, Jr.)

2,000 to 2,500 years, between 15,300 and 12,700 years **B.P.** (before the present). A debate is now under way between proponents of the older concept of a single Spokane Flood and this new interpretation.

Under either concept, a flood episode involved a great wall of water that rushed westward down the Clark Fork valley of Idaho and spread out mostly southwestward across the Columbia lava plains of eastern Washington, creating wide, braided channels (Fig. 8.19). Depths of water were 100 to 200 meters in confined narrows, but in wider, braided **distributaries** it must have been much shallower. Temporary lakes formed along the flood route by ponding upstream from narrow gaps where all the water simply could not get through at once. The largest of these was Lake Pasco, near the city of that name in south-central

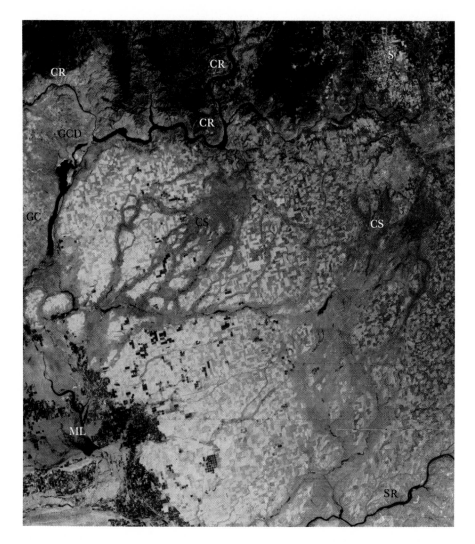

Figure 8.19. This Landsat-1 satellite photo shows the east-central part of the state of Washington. The snakelike black line in the lower right corner is the Snake River (SR). The Columbia River (CR) comes south from the middle upper edge of the photo and is joined from the east by the Spokane River before turning west. The Columbia River widens westward because of Grand Coulee Dam (GCD), which is located at the point where the Columbia abruptly narrows and swings in a wide loop to the north. The large water body extending south of Grand Coulee Dam is impounded by the Dry Falls Dam and fills the Grand Coulee (GC), which was created by waterfall recession during Spokane flooding and by Columbia River erosion when the river was earlier diverted by a glacier lobe. The water bodies and cultivated area in the southwest corner of the photo are in the Moses Lake (ML) region. Extending from northeast to southwest across the area between the Snake and Columbia rivers are a number of interlocking, braided, irregular dark gray streaks. These are the scoured and plucked channels of the Spokane Flood, and they form the channeled scabland terrain (CS). Areas between channels are mantled with fertile loessial soils, and the rectangular checkered appearance is because of wheatfields in different stages of cultivation. The city of Spokane (S) lies near the upper right-hand corner. (Used by permission of Earth Observation Satellite Company [EOSAT], scene ID-8134518145500, July 3, 1973.)

Washington. Floodwaters were nearly 300 meters deep at the narrow eastern entrance to the Columbia River gorge. Water about 130 meters deep, probably in a temporarily impounded lake, is indicated by geological features near Portland.

The flood accomplished much erosion. A cover of loess up to 60 meters thick capping Columbia River lavas was stripped away in all channels, leaving those areas agriculturally sterile. This loess is the parent material for the rich wheatland soils of eastern Washington. Fast-flowing water has a remarkable power of plucking, that is, bodily lifting up and carrying away large, joint-defined blocks of bedrock. The basaltic lavas of eastern Washington could hardly be better suited for plucking because of their nearly vertical **columnar jointing** – a pattern of cracks, formed by cooling, that define polygonal columns. Through plucking, floodwaters created wide, steep-walled channels, waterfalls, and uneven, rough channel floors with closed depressions. Most striking are mesas or tablelands of lava left between braided and interlocked channels. Known locally as **scabs**, they are the source of the term **channeled scablands**, a peculiar landscape (Fig. 8.19).

Some major confined channels developed spectacular waterfalls; one of the greatest (Dry Falls) was more than 5 km wide and 120 to 130 meters high. This abandoned fall sports four large plunge pools, 60 meters deep, at its base. Such falls retreated rapidly, owing to undermining (sapping) at the base, and left in their wake deep, wide, steep-walled gorges called **coulees**, of which Moses Coulee and Grand Coulee are examples.

Among depositional features formed are huge bars of large boulders. Some boulder bars have a succession of wavelike ridges on the surface that resemble giant ripples. They are up to 3 km long and as much as 10 meters high, separated by intervals of 100 meters, and composed of bouldery gravel. Geologists have concluded that they are indeed giant ripple marks built by the flood when it overtopped the bar. The debris-laden torrents also backfilled gravel, sand, and finer sediment into the mouths of tributary channels and valleys, forming **backfill** deposits.

The floodwaters must have carried some large icebergs rich in detritus, for in the Portland area at elevated levels are isolated patches of stony debris including large boulders foreign to the local bedrock. The boulders are too big to have been carried by the floodwaters alone. They needed help, and icebergs could have provided it. These deposits appear to indicate a water depth, probably lacustrine, of about 130 meters near Portland at the time of Spokane flooding.

The Spokane Flood must have been a tremendous spectacle, and it is possible that native peoples were around to witness it, at least for a few

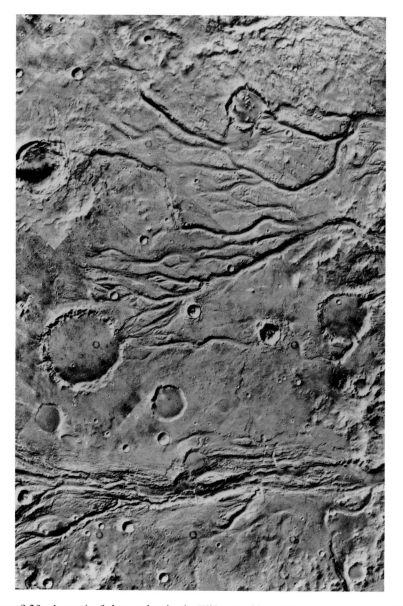

Figure 8.20. A mosaic of photos taken by the Viking I orbiter of Mars during August 1976. The area shown centers at about 17°N, 55°W. North is to the top, and the land surface drops about 3 km in elevation from west to east across this picture. The large crater, center left, is roughly 20 km in diameter. The channels running downhill to the right look like typical terrestrial stream courses. The large complex channel near the bottom is attributed to scouring by a gigantic outburst flood of unknown origin but not unlike the Spokane Flood of east-central Washington State in its effects. (NASA photo.)

brief moments. There is, however, no universally accepted proof that the area was inhabited at the time of flooding.

A fascinating side note is that photographs of Mars taken by Mariner 9 and the Viking orbiters in 1976 show what are interpreted as the scars of similar but even larger floods on the red planet's surface (Fig. 8.20). These martian photographs reawakened interest in Spokane Flood features and stimulated further investigations.

Iceland has repeatedly experienced an unusual type of glacier flood. This country, despite its name, is one of the world's current hot spots – the site of repeated major volcanic activity within historical time. Iceland also harbors much glacier ice, mostly as caps. The combination of glaciers and volcanism yields unusual products. Like big bugs, big hot spots have smaller hot spots upon them, and some of these local hot spots reside beneath Iceland's glaciers. They cause local melting at the glacier's base and sometimes create holes penetrating to the surface. Meltwater accumulating at these spots episodically breaks out from under the ice and floods adjoining outwash plains. These unusual floods bear the totally Icelandic name of **jökulhlaup**. Grimsvötn, a subglacial lake over a hot spot under the Vatnajökul cap of Iceland, discharges a jökulhlaup about every five or six years.

INTERGLACIAL FORESTS. Periods of milder climate, during which ice sheets shrank, possibly even disappeared, are **interglacial intervals**. If they were long enough and the climate sufficiently moderate, dense forests grew in areas vacated by the ice. Subsequently, glaciers reinvaded such areas, knocking down trees, macerating them, and incorporating wood fragments into till laid down over the formerly forested sites. Locally, remnants of the original forest floor are fortuitously preserved beneath the till. Their most arresting feature are still-rooted tree stumps, splintered and worn by the overriding ice. The stumps are reduced to rounded nubs with an asymmetry imparted by ice movement.

One of the best known **interglacial forest** layers is that at the Two Creeks locality of east-central Wisconsin. There the Green Bay sublobe of the Lake Michigan glacier (Fig. 8.5) readvanced over an area occupied by a mature coniferous forest about 12,000 years ago. Remnants of the forest have been thoroughly investigated by paleobotanists, so the climatic and biological environment at Two Creeks just prior to 12,000 years ago is known to be similar to but somewhat cooler and moister than the present. Aspects of a much younger interglacial forest, less than 300 years old, overridden by the Malaspina Glacier on the southern coast of Alaska, are shown in Figure 8.21.

Figure 8.21. Tree trunks, limbs, and other woody debris of an interglacial forest overrun by Malaspina Glacier on the south coast of Alaska less than 300 years ago. This exposure is the wall of a postglacial stream-cut gully. Roots of tree trunks buried by a thin mantle of till project from the slope face near its top and define the position of the original forest floor (FF). The trees were mainly Sitka spruce and larger than most trees currently growing in the area or on the surface of stagnant Malaspina ice (Fig. 1.13).

PERIGLACIAL FEATURES. A host of geological and botanical relationships indicate that climatic conditions along the edge of a continental Pleistocene ice sheet were extremely rigorous and extended for a considerable distance outward over bordering terrains. Conditions and features characterizing this area are termed **periglacial**. Strong winds, abundant water, frequent freeze and thaw, sparse vegetation, and **frozen ground** acting in concert created unusual geological features, so numerous and varied as to deserve a separate book. Indeed, several such books have been published. The only aspects of the periglacial environment to be treated briefly here are perennially frozen ground and **freeze–thaw action**.

Those accustomed to cold climates are familiar with temporary freezing of the ground in winter and its thawing in spring. We remember how wet, soft, and mucky the ground became with the first thawing. That happens because frozen ground contains many segregated grains, spicules, and bodies of ice. The water formed when this ice melts cannot escape by percolation because the still-frozen ground beneath is impervious. It is concentrated and trapped in the thawed surface layer; hence the muddy state.

In many arctic and subarctic regions, especially those of light snowfall, the ground is presently frozen to depths of tens, even hundreds, of meters. This far exceeds the depth of annual thawing (usually a meter or two), so the deeper ground remains frozen year after year. Only the thin surface zone, known as the **active layer**, thaws in summer and refreezes in winter. The ground beneath is **perennially frozen ground** and commonly called **permafrost**.

The annually thawed and refrozen layer is well named, for it is the site of strong forces and marked changes of state. Besides the annual freeze-and-thaw cycle, it experiences other short-term freeze–thaw episodes related to particular meteorological events, especially during spring and fall. Broken water pipes and cracked engine blocks have shown many of us that the expansion of water changing to ice is a powerful force. This same force within the active layer breaks up rocks and mineral particles, segregates stones by size, causes heaving of the surface, and stirs things up. In the thawed state the active layer is highly mobile, owing to abundant water introduced by the ice segregation effects of the freeze–thaw process. Downslope mass flowage occurs readily, rounding and smoothing landscape features and giving them a molded aspect. In some settings coarse rock fragments are segregated into bands extending downslope, forming **stone stripes** (Fig. 8.22). On flat areas the segregated stones outline hexagonal **stone polygons** instead. These structures are known as **patterned ground**.

Winter freezing of an active layer occurs primarily from the top down. Thus, within areas of permafrost a water-rich, mobile mass of still unfrozen material becomes trapped between perennially frozen ground below and an expanding newly frozen layer above. Not only is the unfrozen material squeezed between the two frozen layers, it refreezes unevenly because of differences in grain size as well as variations in surface insulation and **insolation** (warming by the sun). These conditions generate differential stresses within the remaining unfrozen parts, which respond by local contortion. If different layers of material – for instance, light-colored sand and darker silt or clay – are involved, the contortions are easily recognized, even thousands of years later (Fig. 8.23).

In some frozen ground areas, wedge-shaped masses of ice, up to 3 meters wide, extend downward from the surface as much as 10 meters. The internal structure shows that these **ground-ice wedges** have grown by increments, probably annually, and have required decades or centuries to attain their great size. After the climate ameliorates and the ice melts, the form of the wedge may be preserved by an infilling of sand and gravel. These are **fossil ice wedges**, or **ice-wedge casts**.

Figure 8.22. This barren slope, which looks as though it were artifically cultivated by a plow or harrow, is wholly natural. It is the product of strong freeze thaw activity in a region of rigorous climate and perennially frozen ground, the valley of Steele Creek, Yukon Territory, Canada. The larger stones within the broken rock debris mantling the surface are segregated from the finer debris and concentrated into stripes extending directly downslope by the freezing and thawing, accompanied by concurrent downslope mass creep.

People living in temperate areas of midwestern United States or central Europe may know that regions far to the north are underlain by perennially frozen ground. Probably few stop to think that, perhaps, during the Ice Age their region experienced periglacial conditions, including perennially frozen ground. The evidence that it did is provided by patterned ground, contorted layers, and ice-wedge casts, among other phenomena.

RECAPITULATION

People living near the edge of areas inundated by continental ice sheets, or in the lower part of a glaciated canyon, are well acquainted with glacial deposits. They consist of rock debris transported, often over considerable distances, by ice before being deposited directly as till or given to streams of meltwater for further transport and deposition as glaciofluvial material.

Depositional features most commonly seen include moraines, erratics,

Figure 8.23. Complex deformation of once-horizontal sand (light) and silt (dark) layers, shown in this photograph of Pleistocene glaciofluvial deposits in central Illinois, is the result of differential squeezing of unfrozen materials between underlying perennially frozen ground and an overlying annually frozen layer during the Ice Age. This occurred repeatedly in areas peripheral to the Pleistocene continental ice sheets in both North America and Europe.

outwash plains and trains, ground-moraine sheets, drumlins, and various ice-contact features, such as kettle holes, kames, and eskers. Glacioeolian deposits, derived by wind from outwash plains and river flood plains, make up sand-dune sheets and mantles of loess (dust). Moraine-dammed lakes enhance the scenery.

Glacial-lake deposits containing sedimentary couplets of clay and sandy silt (varves) are valued for measuring geological time. Some areas, northeast and east-central Washington State, for example, bear the scars of huge, catastrophic floods generated by the release of waters impounded behind dams of glacial ice.

Until now we have focused largely on glaciers, how they work and what they do. We next give attention to the causes of glacial episodes and the history of waxing and waning ice masses during the past million years. We will also concern ourselves with the impact of possible glacier fluctuations on the future of the human species.

9

The past and the future

The preceding discussions of glaciers and their work can now be related historically to ice ages, their nature and possible causes, and conditions and events associated with them.

The most recent Ice Age involved the past 1.0 to 1.5 million years (the **Pleistocene Epoch** of geologic time), a period when continents of the Northern Hemisphere were partly covered by sheets of ice. The terms *Ice Age* and *Pleistocene* are often used interchangeably, although they may not be strictly equivalent. We don't know yet whether the Ice Age is over, but by definition, the Pleistocene ended 10,000 to 11,000 years ago, giving way to the **Holocene Epoch**.

The Pleistocene Ice Age was not a unique event; ice ages have occurred earlier in Earth's history, almost certainly as long ago as 2 billion years and perhaps earlier. This is important in considering the causes of glacial periods. We are fortunate that the last Ice Age took place so recently. Its record is still well preserved, so we know a lot about the conditions that characterize such events. This knowledge leads to an interesting speculation that is of major concern to all humankind: The last great Ice Age may not be over. We could currently be living in the early part of a relatively mild interglacial interval, and the ice may come back.

THE CONSTITUTION OF AN ICE AGE

Before further considering ice ages past, present, or future, we should develop some concept of just what an ice age encompasses. One learned tome states that an ice age is a time of extensive glacial activity, which is true but does not indicate how extensive or where the glaciers were. The dictionary unhelpfully says an ice age is a glacial epoch but does not define *glacial epoch*. A highly regarded encyclopedia equates an ice age to the Pleistocene Epoch, which may be close to the truth but is not very enlightening. The encyclopedia article, however, does mention that continental ice sheets characterize an ice age – a thought with merit.

Let us frame a definition as follows. An **ice age** is characterized by conditions that, from time to time, cause ice sheets approaching 1 million km² or more in area, and a maximum thickness measured in thousands of meters, to develop on nonpolar continents. Only two sheets with these dimensions exist today, one on the Antarctic Continent and one on the subcontinent of Greenland. The Antarctic Continent is unquestionably polar, and Greenland qualifies too if we believe airline companies that fly over it. These ice sheets do not, however, qualify the present as an ice age according to our definition. We purposely emphasize continental ice sheets because scattered ice streams and ice caps in high-standing mountain areas within nonpolar regions are not evidence of ice-age conditions; witness the present situation.

Our definition says "from time to time" because during an ice age there may have been intervals when the continental ice sheets shrank dramatically and possibly disappeared, even though a relatively cool environment prevailed. Such periods are known as interglacial intervals, but interglacials are properly parts of an ice age because the glaciers return. Later we will address the question whether the period in which we live is an interglacial.

A definition of ice-age conditions depends strongly upon one's location. It may be difficult to convince a native Greenlander or someone just returned from Antarctica that the Ice Age is over. Even though the Antarctic Ice Sheet may have been 10 percent larger 20,000 years ago, and the Greenland Ice Sheet perhaps 25 percent larger, glacial conditions still prevail in both regions. That is why the term *nonpolar* in our definition is important. For the vast bulk of the world's population, the present environment is not one of glacial conditions. They have either seen the end of the last Ice Age or are living through an interglacial interval within it. Only time will tell which.

CONTINENTAL SHEETS OF THE LAST ICE AGE

It will be helpful in understanding former continental glaciers to refer occasionally to the features and characteristics of the existing continental sheets on Antarctica and Greenland (Fig. 9.1).

Let us consider those continental sheets that thrived during the Ice

Figure 9.1. This vast, smooth desert of ice is what the continental ice sheets of North America and Europe must have looked like in countless places and times during the Pleistocene. In this instance, we are looking at the surface of the Greenland ice sheet about 400 km from its western edge. The ice is thick enough to obscure completely any relief of the underlying bedrock terrain, and conditions are cold enough so there is little wastage of snow, except locally by wind drifting. The lack of landmarks makes this an easy place to get lost, especially in a fog, so this expedition included an astronomer to do the navigating. (Photo courtesy of Carl S. Benson.)

Age but are now completely gone, at least for the present. There were only two: one in North America and the Eurasian ice sheet. Both were compound in the sense that when full-grown they consisted of bodies that began in separate centers and spread until integrated into a continuous mass. No southern hemisphere continent, other than Antarctica, harbored an Ice Age sheet; they were all too equatorial. Only local highland glaciers existed on other southern continents.

The North American Ice Sheet covered essentially all of Canada, many of the Canadian arctic islands such as Ellesmere and Baffin, some of Alaska, and the central United States. At maximum it may have joined at its northeastern corner with the ice sheet of Greenland (Fig. 9.2). (The Greenland sheet is usually not treated as part of the North American Ice Sheet for the same reasons that Greenland is not usually considered part of North America.) The North American sheet consisted of three parts: a large core body covering central and eastern Canada, the *Laurentide Ice Sheet*; a western part centered over the mountains, the *Cor-*

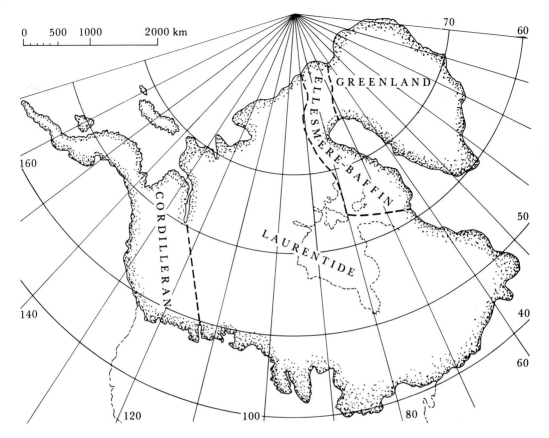

Figure 9.2. Map of area covered by the Pleistocene North American ice sheet, a composite of all advances. The three principal components – Laurentide, Cordilleran, and Ellesmere-Baffin – are identified. The Greenland Ice Sheet is not regarded as part of the North American complex, although they may at one time have joined. (After G.H. Denton and T.J. Hughes, eds., 1981, The Last Great Ice Sheets, *Fig. 6.15, New York: John Wiley & Sons, 484 pp.)*

dilleran Ice Sheet; and a smaller northeastern appendage, the *Ellesmere–Baffin glacier complex*. At maximum the North American sheet covered about 16.2 million km² of land and extended south into midwestern United States almost to the junction of the Ohio and Mississippi rivers. By comparison, the antarctic sheet covered 15.3 million km² at maximum.

Sheets of this dimension are generally thick enough to submerge most of the underlying topography. This was particularly true of the Laurentide sheet, which occupied an area of modest relief in central interior Canada. It was a vast plain of snow and ice above which a few broad ice domes rose with slopes so subtle that anyone climbing them would probably have had trouble deciding when the summit had been reached. The Laurentide sheet resembled the ice currently covering East Antarctica, which completely overwhelms the underlying topography. The Cordilleran and Ellesmere–Baffin masses were more like the West Antarctic

173

sheet, which largely but not completely buries an island archipelago and rough, mountainous terrain, so there are many projecting islands of bedrock (nunataks) and mountain ranges rising above the surrounding ice.

Geophysical soundings show the Antarctic Ice Sheet to have a maximum thickness of more than 4,776 meters. The weight of such thick ice can depress the underlying land so much that it comes to lie below sea level. This has happened in parts of Antarctica and under one-third of Greenland. From the amount of depression caused in central Canada by the Laurentide sheet and other considerations, its maximum thickness is estimated to have been more than 4,950 meters. Following disappearance of this ice, the land has been slowly rebounding, but in places it still has not yet fully recovered. Hudson Bay, for example, will cease to exist as a significant body of seawater when **glacial rebound** is complete. It is largely a temporary relic of the depression of central Canada by the North American Ice Sheet.

If one looks at a map of midwestern United States with a line showing the maximum position attained by the North American Ice Sheet during the last Ice Age (Fig. 9.3), it is easy to form a mental image of a relatively smooth, regular ice edge extending from Montana southeastward about to the junction of the Ohio and Mississippi rivers, and then northeastward toward New England. Had we been on the spot at any time during the Ice Age, we would have seen that the edge of the ice sheet was actually highly irregular and strongly lobate. Additionally, the various **ice lobes** did not act in concert; some were at advanced positions while others were still advancing or receding. The ice-edge was constantly changing, although slowly.

Much thought has been given to where and how the North American Ice Sheet originated and to its subsequent growth and development. A favored concept is that ice streams and ice caps – like those existing today on the large, high northeast Canadian arctic islands (Baffin, Ellesmere, and Devon, among others) – expanded, coalesced, and joined with an independent center of ice development in the nearby relatively elevated Ungava and Labrador plateaus of mainland Canada. This formed a protolaurentide sheet that grew and flowed southwestward into central interior Canada, eventually becoming thickest over Hudson Bay. At about the same time, similar developments in the western mountains were creating the Cordilleran Ice Sheet. The two bodies eventually merged not far east of the mountains to make a huge, integrated sheet (Fig. 9.2).

Once an ice sheet reaches a critical size, it is to some degree self-perpetuating. It strongly influences its own weather and the weather of the surrounding country. Ice over Hudson Bay was mostly about 4,500 meters thick. Even though the land was depressed by this heavy burden,

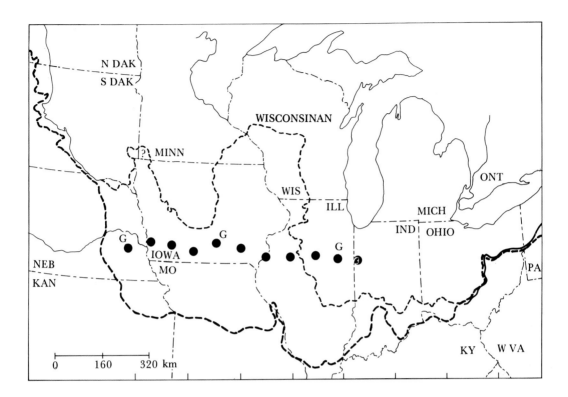

Figure 9.3. Map showing the southern limit of the Pleistocene glaciated area (heavy dashed line) of central United States and the boundary of the last (Wisconsinan) ice advance (light dashed line). The irregular, heavy-dot line (G) represents the course of the imaginary gorge discussed in the text. (After R.F. Flint, 1971, Glacial and Quaternary Geology, *New York: John Wiley & Sons, p. 545.)*

elevation of the ice surface was at least 3,000 meters above sea level. If we placed a large mountain mass of that height in central Canada today, it would almost certainly develop glaciers within a short time.

A side note of interest is that some areas that seemingly should have been ice-covered were not. Interior Alaska is an example (Fig. 9.2). Its modestly elevated Yukon Plateau lay west of the northwest corner of the Laurentide sheet, south of a large ice mass on the Brooks Range, and north of an extensive glacier complex in the Alaska Range. It was almost fully enclosed by large ice accumulations on three sides, but it contained no significant amount of glacier ice. The environment was simply too dry.

Like the North American sheet, the Eurasian body formed in three principal areas and was initially nourished by them. By far the dominant center was the highland area of Scandinavia, which generated the large Scandinavian Ice Sheet. At maximum that sheet merged with a separate glacial complex to the southwest mantling the British Isles and with another, smaller sheet well to the east near the Ural Mountains. This last

175

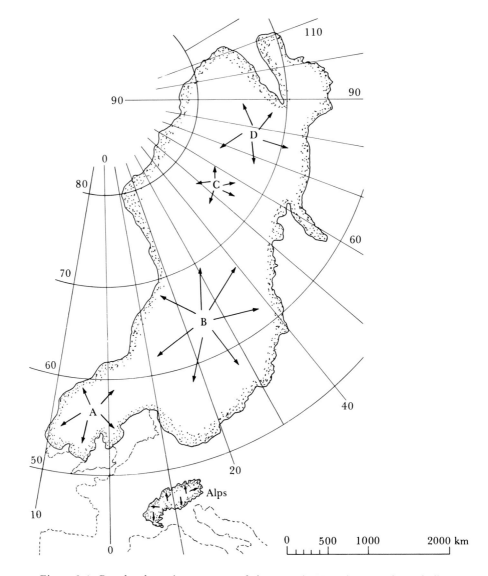

Figure 9.4. Postulated maximum extent of the composite Eurasian Ice Sheet of all stages showing principal areas of accumulation: British Islands (A), Scandinavian highlands (B), Ural Mountains (C), upland areas of northwestern Siberia (D). (After G.H. Denton and T.J. Hughes, eds., 1981, The Last Great Ice Sheets, Fig. 6.15, New York: John Wiley & Sons, 484 pp.)

body formed in those mountains and in adjacent upland areas of north-western Siberia (Fig. 9.4). The combined Eurasian Ice Sheet was about 60 percent as large as the North American sheet, covering roughly 9.4 million km². The bulk of its ice, 70 percent, was in the Scandinavian sheet, which overran the Baltic region, Poland, and much of Russia, extending south into the central European plain. It fell well short of joining with ice flowing north out of the Alps. The maximum thickness of this sheet has been estimated at about 3,300 meters. In central Europe, its southern edge was more regular than the edge of the North American Ice Sheet, but in Russia the wide valleys of the Dnepr (Dnieper) and Don rivers caused strong lobation.

From its harsh climate one might have expected Siberia to generate a huge ice sheet, but it did not. Aside from the relatively small sheet of roughly 2.7 million km² in its northwest corner, classed as part of the Eurasian body, Siberia had less than a dozen scattered and separate glacial complexes on mountainous uplands. Their cumulative area amounted to only 700,000 km². The reason was lack of moisture, just as in central Alaska.

We do not have reliable information on the rate at which the continental ice sheets of the last Ice Age grew. The time between the first invasion of midwestern United States by lobes from the Laurentide sheet and the initial formation of ice on the Ungava–Labrador plateaus may have been at least 20,000 to 30,000 years. This determination is difficult to make because an ice sheet destroys its own growth record as it expands and advances. The pattern of deglaciation for the last glacial episode, from the most advanced position in the midwestern United States to virtual disappearance of the continental sheet ice from North America, is still being investigated and debated. It involves about 12,000 years, starting 18,000 years ago and extending to a small residual ice cap, now gone, in central Labrador about 6,000 years ago.

GLACIAL AND INTERGLACIAL HISTORY OF THE ICE AGE

The Pleistocene Ice Age was not a simple cycle of growth, stabilization, recession, and disappearance of ice sheets; rather, it involved a complex succession of events. The fragmentary record of this history preserved on land indicates that both the North American and Eurasian sheets went through at least four cycles of formation, expansion, recession, and probable disappearance within the Ice Age. In no instance did these events proceed smoothly and regularly; advances were irregular and erratic and recessions likewise, three steps one way and two steps the other. This behavior was controlled largely by variations in meteorological and climatic factors, which are notoriously unruly and irregular phenomena, both short- and long-term. The record of variations in ocean temperatures, recorded by differences in oxygen-isotope ratios (^{18}O:^{16}O) within the shells of single-celled organisms preserved in sediments on the

177

seafloor, permits an interpretation that glacial episodes were more numerous and variable than the land record suggests.

On land, the most useful record of Ice-Age history is preserved in the glacial drift left by ice sheets. This includes not only till deposited directly by the ice, but also the outwash and loess carried well beyond the ice margin. It is now recognized that the four episodes of expanded glacial activity identified on land were separated by intervals of nonglacial warmer and sometimes drier conditions. These interglacial intervals lasted as long as or longer than the glacial episodes. In turning to ice-sheet deposits to learn the history of the Ice Age, let us focus on the record of the Laurentide Ice Sheet, because the simplicity of terrain makes it the easiest to understand.

Imagine that a good-sized river, flowing from west to east, had cut a steep-walled gorge or canyon about 100 meters deep and 800 km long on a winding course starting in eastern Nebraska, extending across southern Iowa, and continuing across central Illinois to the Indiana border (Fig. 9.3). Pleistocene glacial deposits would be nicely exposed in the upper part of the steep walls of this gorge all along its length.

Within the mantle of glacial drift we could expect to see distinct layers of three principal types of material: glacial till, glaciofluvial sands and gravel, and loess. One other useful feature would be zones of weathered material at the upper boundaries of some of these layers. The weathered zones could be recognized by brownish color, richness in clay, distinctive structures such as clods and columns, and generally soily characteristics. In Nebraska and Iowa we would see that a lower assemblage of till, glaciofluvial outwash, and loess units was separated from an overlying assemblage of similar materials by one of these deeply weathered zones.

The basal unit of the lower sequence would be a glacial till with a clay-rich matrix and a dark, dirty appearance imparted by its content of weathered rock debris and ground-up woody material. We might collect some wood fragments and gleefully rush off to the nearest carbon-14 or radiocarbon dating laboratory, hoping to find out how old the lowermost till layer is. We would be doomed to disappointment: The laboratory would tell us the wood was radioactively dead – too old to give a radiocarbon age. The carbon-14 method cannot measure ages greater than 50,000 to 70,000 years. The coherence and compactness of this till indicates a considerably greater antiquity, as does its position at the bottom of the Pleistocene sequence. It may be as much as a million years old.

Nowhere does this old till compose the surface layer; it is always buried under younger deposits. Geologists have elected to call it the *Nebraskan till*, a product of the **Nebraskan glaciation**, because of exposures long ago identified in the eastern part of that state. The weathered nature of much of the rock debris and the rich supply of macerated plant material in this till is attributed to the fact that Nebraskan ice was the first to invade the region. The preglacial surface offered a greater amount of such material to the glacier than to later ice invasions.

The deeply weathered clay-rich zone, in places 3 meters thick, at the top of the Nebraskan drift is appropriately termed **gumbotil**. Gumbotil is created by long-continued **weathering** under warm, moist conditions, and a 3-meter thickness probably required many tens of thousands of years. Such gumbotils are interpreted to be the product of long intervals between episodes of glaciation. During these interglacials, the North American sheet not only receded but probably disappeared entirely.

The upper till, exposed near the top of our gorge and lying above the lower gumbotil, is a widespread deposit that, along with associated glaciofluvial materials and loess, covers most of eastern Nebraska, northeastern Kansas, northern Missouri, and about two-thirds of Iowa. It, too, must be of some antiquity, because a second gumbotil horizon, up to 4 meters thick, is preserved in places at its top. The second interglacial interval, implied by this gumbotil, was possibly longer and warmer than the first. The glacial drift upon which it has been developed has long been attributed to the **Kansan glaciation**, so named because of unusually good exposures in the northeastern corner of that state.

To see additional glacial units, we have to go east almost to the Iowa–Illinois border. There we will encounter a much younger-looking layer of drift consisting, in different exposures, of till, outwash, or loess resting on top of Kansan deposits. This drift covers the surface of western and southern Illinois almost to the middle of that state. Its topographic expression, displaying modest but recognizable vestiges of moraines and kames, differs from the subdued older Kansan mantle seen in Iowa. The till is less consolidated and has fresher rock fragments and a much less deeply weathered surficial layer, only a meter or so thick. The designations *Illinoian drift* and **Illinoian glaciation** seem appropriate for these deposits.

The interglacial interval that followed had a highly varied flora and fauna, featuring large mammals, such as bison, bears, mammoths, and ground sloths, in great numbers. This would have been an interesting interglacial in which to live. At least, we would have had something to eat.

If we continued eastward, suddenly in the middle of Illinois we would come upon a still younger layer of drift resting upon the Illinoian deposits. This new drift would look remarkably fresh, with only modest manifestations of weathering on its surface. It would be characterized

Table 9.1. *Glaciations in North America and Europe*

North America	Northern Europe	Alps	British Isles
Wisconsinan	Weichselian	Würm	Newer Drift
Illinoian	Saale	Riss	Gipping
Kansan	Elster	Mindle	Lowestoft
Nebraskan	—	Günz	Weybourne Crag

by a different type of landscape featuring moraines, a rolling up-and-down topography, many closed depressions, and few well-formed lines or patterns of drainage or dissection. This is the realm of the last of the four Pleistocene stages, named the **Wisconsinan glaciation** because of features in that state. We have entered the area from which the North American sheet began its last, dramatically rapid recession about 18,000 years ago. The last small vestiges of the sheet are thought to have disappeared completely somewhat less than 6,000 years ago.

Thus, in middle North America four distinct phases of Pleistocene glaciation, separated by interglacial intervals of warmer and usually drier climate, are recognized from the deposits left by the Laurentide Ice Sheet. From oldest to youngest, these are named Nebraskan, Kansan, Illinoian, and Wisconsinan. The interglacial intervals also have names, and if we include the present Holocene Epoch as part of an interglacial, they also number four. Other less extensive episodes of glaciation probably occurred during the Pleistocene Ice Age, but the evidence for them in most places on land has been so modified, obscured, or destroyed that they go unrecognized. Iceland, with ten possible episodes of glaciation is an exception, and seafloor sediments record many more than just four cold periods during the Pleistocene.

European glaciers seemingly experienced a similar four-phase history during the Pleistocene (Table 9.1). The names, drawn from geographic localities, are naturally different, and there is limited proof of any across-the-board correlation of the four phases on opposite sides of the Atlantic. Radiocarbon dating within deposits formed during the past 50,000 years suggests that the North American and Eurasian sheets behaved similarly over that interval. Unfortunately, we lack accurate dating techniques for correlating older Pleistocene deposits unless they happen to be interlayered with volcanic materials suitable for **potassium-argon dating**.

As shown in Table 9.1, glaciers in the Alps have followed a behavioral pattern similar to that of the ice sheets, and the same appears to have been at least partly true for mountain glaciers of western United States. More importantly, mountain glaciers in the southern hemisphere seem to have fluctuated in concert with glaciers of the northern hemisphere, insofar as dating by ^{14}C can show. This is important in considering the causes of and variations within an ice age.

ANCIENT GLACIATIONS AND ICE AGES

Identifying ancient glaciations is not easy. In Chapter 7 the difficulty of distinguishing glacial till from similar deposits created by nonglacial processes, particularly debris flows, was discussed. That difficulty is compounded by **lithification** of the deposits, by their scattered and fragmentary nature, and by alteration of the geographic setting in which they were created. Nonetheless, experienced geologists are convinced they have found many examples of true **tillite** (lithified till) in the geological column going back more than 2 billion years. They have satisfied at least some of their critical colleagues that the evidence and their interpretations are valid, especially in those fortuitous instances in which the tillite rests upon glacially scoured and striated bedrock surfaces.

Let us regard as fact that there have been a number of glacial episodes in Earth's history; the oldest so far recognized occurred about 2.2 billion years ago. Many exposures of tillite 600 million to 700 hundred million years old have been identified on several continents, and more localized occurrences have been demonstrated in rocks ranging from 450 million to 250 million years in age in various parts of Africa, South America, and Australia. Earth may have experienced many additional episodes of glaciation that go unrecognized, sometimes because the evidence may have been destroyed. Preservation and discovery of the evidence are both largely matters of luck. Chances are good that, were we able to return in 100 million years to those parts of North America inundated by the great Pleistocene ice sheet, we would be unable to find convincing evidence that the area had ever been glaciated.

Some, perhaps most, tillites may not represent a Pleistocene-type ice age. The reason why can best be illustrated by an imaginary exercise. Under present environmental conditions Antarctica shows that a polar continent will be covered by a great sheet of ice. (A continental mass, if there were one, at the north pole would almost surely be similarly inundated.) Now play a little game of chess by replacing Antarctica with Australia. In due time Australia would have an ice sheet and undergo severe glaciation. If Australia then moved back to its present position, the ice sheet would melt, but for some time thereafter evidence of erosional and depositional products of glaciation would be recognizable.

That is the problem with respect to many ancient glaciations: Are they the product of a worldwide ice age or the result of a certain continental land area occupying a polar position long enough to become glaciated? Modern concepts of plate tectonics, continental drift, and polar wandering would permit different parts of existing continents, particularly in the southern hemisphere, to have occupied a polar position sometime in the past. Several of the glacial episodes recognized within the past 450 million to 250 million years are probably polar glaciations, not the result of ice ages.

It might be argued that ice-age conditions are required for even a polar continent to experience glaciation. That is debatable, for we know the Antarctic Continent had a large ice mass 13 million years ago and at least some ice as much as 37 million years ago, when the rest of Earth was experiencing anything but ice-age conditions. We also know that from 600 million to 250 million years ago, the average temperature on the planet was generally about eight degrees Celsius warmer than at present, suggesting that any glaciation during that interval was probably polar.

The question of how to distinguish products of a polar glaciation from those of a true ice age thus arises. One possible way would be to show that a large number of bona fide tillite exposures, widely distributed in different continents, have a closely corresponding age. It is unlikely that all of them could have been in a polar position at the same time. Other evidences of widespread ice-age conditions existing concurrently would be supportive.

The two best candidates for true ice-age status among the many episodes of ancient glaciation are one occurring within the 600-million to 700-million-year B.P. interval and one roughly 230 million to 250 million years B.P. Many geologists would probably now accept at least those two glaciations as indicators of true ice ages. Chances are good that there have been others for which the evidence has been obscured. Whatever the cause of an ice age, it was not something unique to the Pleistocene Epoch.

THE CAUSE AND CONTROL OF ICE AGES

It is desirable to have in mind the difference between what brings an ice age into being – that is, creates ice-age conditions – and what factors cause variations during an ice age, such as advances and recessions or the change from glacial to interglacial conditions. The cause of an ice age is clearly climatic change, but what sort of change and how great a change was it, and most importantly, what caused the change?

The welter of proposals regarding causes of ice-age conditions resembles a bucket of worms. Some order can be brought to this complex by recognizing the following classes of causes: those having to do with terrestrial processes and influences, those based on astrophysical or Earth-orbital considerations, and those that come to Earth from extraterrestrial space. Before addressing these categories, the magnitude and type of climatic change needed to start an ice age are worth discussing.

The climatic change needed to initiate an ice age

Weather and climate are the result of a complex interplay among many independent and dependently variable meteorological factors. Changing one can cause changes in many others. We learned from creating our own little ice stream that an important virtue in making glaciers is thrift in conserving snow. Accordingly, we understand that changes in climate resulting in more accumulation of snow in the right places on nonpolar continental masses are the sort of thing needed to create an ice age.

Variations in meteorological factors causing most climatic changes of the past are largely a matter of guesswork, and with respect to the Pleistocene Ice Age they are speculative. The meteorological factor that most likely comes to mind first is temperature. Fortunately, there are some data on temperature changes, so let us focus attention upon temperature, recognizing that temperature in turn influences other factors and conditions that play a role in helping to create an ice age.

The temperature change needed to initiate an ice age depends upon the base from which one starts. Present conditions may not provide a realistic datum, because we may be living in an interglacial interval within an ice age, and the change required to initiate renewed glaciation could be only a fraction of that required to start an ice age from scratch. The fossil record suggests that during fully 90 percent of the last 600 million years, the average worldwide temperature has been about 22° C (72° F). That figure is based largely upon plant and animal remains found in old rocks, at best a fragmentary record subject to different interpretations. The current worldwide average temperature is about 14° C (58° F), so the present represents a relatively cool phase of world climate. This cooling of about eight degrees occurred gradually within the last 37 million years and is not something that suddenly and precipitously developed coincidentally with the last Ice Age.

We do not know, with any degree of accuracy, what decrease in temperature would be required to initiate renewed ice-sheet formation on northern hemisphere continents under current conditions. One experienced investigator calculates that glaciers would re-form and grow in the plateau areas of Labrador and Ungava, Canada, if mean temperatures there fell by about three degrees Celsius. Adding those three degrees to the eight-degree difference between the present average world temperature and the average prevailing over much of pre-Pleistocene time suggests that a change of about eleven or twelve degrees was required to initiate the Pleistocene Ice Age. Temperatures could have dropped farther during the Ice Age, of course.

Not everyone accepts the hypothesis that an ice age requires colder temperatures. Currently, the Antarctic Ice Sheet might be larger, if Antarctica were warmer. The reasoning is that Antarctica is presently colder than necessary to preserve all its snow. If air masses nourishing the ice sheet were warmer, they could contain more moisture and hence deliver more snow without increasing ablation, hence enabling the Antarctic Ice Sheet to expand. Russian and French glaciologists have indeed shown that the East Antarctic Ice Sheet accumulated more snow during an interglacial interval than during glacial periods.

The same reasoning could possibly apply to central Canada, suggesting that what the area needs to form glaciers right now is more snow, not colder weather, and that somewhat warmer air masses moving into the region could provide that additional snow. This is probably not a sound argument, however, since central Canada already receives more than enough snow to make glaciers; it just does not save the snow it already gets. There is adequate evidence from so many sources showing that temperatures during glacial phases of the last Ice Age were actually colder than the present that it seems reasonable to extrapolate to a requirement of colder conditions to initiate an ice age.

Two other concepts bearing on ice-age origins need to be noted. First, owing to feedback mechanisms, an ice age, once it starts, can be self-enhancing. The feedback occurs because a large glacier creates a climatic environment – involving temperature, winds, cloudiness, air circulation, and other factors – that is favorable to glaciers. One investigator remarked that it is easier to start an ice age than to stop it. A second concern is the matter of triggering processes, little changes that unleash larger developments. They may not be the basic causes, but without triggering, perhaps nothing would happen. Triggering, however, may be more important in controlling events occurring within an ice age than in setting the ice age under way, as we will see.

Terrestrial factors and influences

Terrestrial factors possibly contributing to development of an ice age can be subdivided into three classes: those involving the solid body of Earth, those involving the hydrosphere, and those involving the atmosphere.

Among solid-earth processes, **tectonic activity** (including plate tectonics) and volcanism are two of the more important. Nearly all treatments of ice-age development recognize that the condition known as continentality is important. Land masses of continental proportions, the higher the better and in proper position, favor the formation of local glaciers and their expansion into continental ice sheets, the trademark of any ice age. **Plate tectonics** can move continents into the right places and, with the help of associated tectonic activity, can create highland areas. In the context of Earth's history, the present appears to be a time of marked continentality with emergent, high-standing continents. These continents have risen an average of about 600 meters over the past 10 million to 15 million years. Fortunately for glaciers, continents of the northern hemisphere were in the right place to form the Pleistocene ice sheets; those of the southern hemisphere, except for Antarctica, were not.

Volcanism can have either a positive or negative influence on glaciation. Volcanic gases, such as carbon dioxide and water vapor, emitted into the atmosphere inhibit the transmission of outgoing radiation from Earth's surface, contributing to a greenhouse type of warming, generally regarded as a negative influence. Explosive volcanism, injecting fine particulate material high into the atmosphere, is generally regarded as a positive influence, because the particles act as crystallization nuclei for snowflakes, and they shield Earth from incoming solar radiation, promoting cooling and reducing wastage of snow.

The problem with volcanism as the primary cause of an ice age is the large amount and long duration (tens of thousands of years) of explosive activity required. Earth has undoubtedly experienced episodes of volcanism of even greater duration, but much of the activity has been non-explosive. We may be living in a period of relatively high explosive activity right now, but it is not creating an ice age, although it does affect the climatic environment to a minor degree. Earth has seemingly experienced many episodes of great volcanic activity without generating an ice age in each, if any, instance. A one-to-one relationship of volcanism and glaciation has not been demonstrated from the geological record.

In considering the role of the hydrosphere, attention focuses on the oceans. Glaciation is a climatic extreme or aberration, a departure from the norm. The ocean is the great moderator, and its influence is largely the opposite of continentality, which favors glaciation. Nonetheless, the ocean as the source of the moisture that makes glaciers is an indispensable contributor to an ice age.

Two changes occurring in oceans that could strongly influence glacier development are shifts in ocean currents and changes in sea-ice cover in polar regions. Imagine what a welcome source of moisture an open, ice-free Arctic Ocean would be to all cold land masses surrounding it – Siberia for one. This factor has been carefully considered in a model proposed for causing the repeated shifts from glacial to interglacial conditions during the last Ice Age. A major stumbling block is a mechanism for freeing the Arctic Ocean of its cover of sea ice. Warm ocean currents from the Atlantic Ocean is one possibility considered. This is a way ocean currents might influence the formation of glaciers.

The atmosphere is the great communicator and the workhorse of any ice age. It delivers the goods and plays a major role in controlling the wastage that ultimately destroys ice. It is the most mobile, flexible, and changeable of the major factors. Changes in atmospheric composition, shifts in wind patterns, storm tracks, semipermanent centers of high and low pressure, and the jet streams could all bear on glacier formation. Solar radiation is a major factor in the wastage of ice, and it has to pass through the atmosphere. Changes in atmospheric composition affect the flux of incoming solar radiation, and they can play an even greater role in controlling outgoing terrestrial radiation. Increases in carbon dioxide and water vapor produce a warming influence by trapping outgoing long-wave heat rays.

From these ruminations one might reasonably conclude that terrestrial factors of all three categories can play an important role in influencing the course of an ice age but that, taken by themselves, they may not be able to start one. Many students of the problem would agree. The terrestrial factors appear to need help, so let us see whence that help might come.

Astrophysical considerations

Three periodic variations in Earth's orbital behavior are thought by many people to significantly control events within an ice age, possibly even to trigger its initiation by capitalizing upon conditions governed by other factors. In themselves, these orbital variations do not appear adequate to cause an ice age; otherwise Earth would have experienced glaciations periodically throughout its history.

The first orbital variation involves changes in tilt of Earth's rotational axis with respect to its orbital plane around the sun. The tilt is usually expressed as a divergence from perpendicular to the orbital plane, and is currently 23°27′. Over time intervals between 38,000 and 45,000 years, the tilt moves back and forth between 21°39′ and 24°36′. In a sense, the axis wobbles a little (2°57′) over a long interval of time.

Earth's orbit around the sun is not a perfect circle but an ellipse. All ellipses have two "centers," called **foci**, and the separation between them is a measure of the **eccentricity** of the ellipse, becoming smaller as the ellipse more closely approaches a circle. The sun lies at one focus of Earth's elliptical orbit, so the distance between the two bodies changes as Earth makes its annual journey. The eccentricity of Earth's orbit is variable over an interval of about 92,000 years. This affects the amount of radiation received by Earth at **perihelion**, the point in its orbit closest to the sun, and at **aphelion**, where it is farthest from the sun.

The third orbital variation involves precession of the equinoxes. An **equinox** occurs whenever the sun is directly above the equator, and the tilt of Earth's axis causes this to happen twice a year as Earth orbits the sun. The two equinoxes occur at positions on Earth's orbit that shift or precess slowly over an interval ranging from 16,000 to 26,000 years. This in turn affects the time of year at which Earth is at perihelion. Currently, perihelion is about January 2, which means that winters in the northern hemisphere are a bit warmer than average and summers in the southern hemisphere a bit cooler. About 9300 B.C., perihelion was on June 21, and these temperature relationships were reversed.

These orbital variations do not affect the total solar radiation received by Earth, but they do change its distribution in ways that are expressed in seasonal differences of climate. Variations in distribution of solar radiation for different latitudes in both hemispheres have been calculated and recalculated several times, as far back as 600,000 years. Fluctuations of the resulting radiation curves are irregular in amplitude and wavelength. They are not exactly opposed in opposite hemispheres, as one might expect, but they are not in phase. The curves of greatest interest are those for 60° north latitude, close to the center of the large Pleistocene ice sheets. Computer modeling suggests that temperature changes resulting from orbital variations are less than two degrees Celsius at this latitude. This hardly seems great enough to serve as anything more than a triggering mechanism. Certainly, it is not large enough to start an ice age except indirectly, by stimulating some other process to do the job.

It is difficult to make a convincing case for correlation of radiation curves with glacier fluctuations on land. Changes in ocean temperatures, recorded largely by oxygen-isotope ratios in corals along seashores and in small animal shells in seafloor sediments seem to correlate best with the curves. The oceanic realm does a much better job of preserving a detailed and complete record, and where the oceanic and terrestrial records are not compatible, the oceanic sequence is likely to be the more reliable.

It looks as though orbital variations can best serve as a triggering mechanism. That does not mean their contribution is trivial; they are there and cannot be ignored.

Extraterrestrial factors

The world's climatic system is driven by solar energy, so it is natural we should look to old Sol, among other possible extraterrestrial influences, as the cause of an ice age. Within the time of human measurements, luminosity of the sun has varied so little that we speak of a **solar constant** of radiation as received at the outer surface of Earth's atmosphere. Recent measurements, however, suggest a slow, very small decrease in solar luminosity. To initiate and sustain an ice age, a significant change in solar radiation is needed, in amount or character or both. A shift to less long-wave, infrared radiation and more short-wave, ultraviolet radiation would be the sort of change in character that might promote cooling on Earth.

Changes in solar radiation could involve events occurring within the nuclear furnace of the sun or anything that could impede or alter transmission of energy from sun to Earth. Interstellar space contains large quantities of material, such as cosmic dust and veils and globs of gaseous substances. Our solar system passes through such masses from time to time on its journey through space. Such a passage, easily requiring a million years, could perturb the sun's luminosity by interfering with transmission of radiation or by creating disturbances on or within the sun through infall of a substance that becomes involved in the sun's internal reactions.

All such musings are rank speculation, but speculation in which the sky is the starting point rather than the limit. The thought that an ice age may be born in the heavens appeals because of its mysticism and beauty. That may not be the way to do science, but in this instance it's nice to be able to appeal to something far out. A change in solar radiation might not launch an ice age unless Earth were ready in other respects, specifically the terrestrial factors earlier noted. Many influences may be

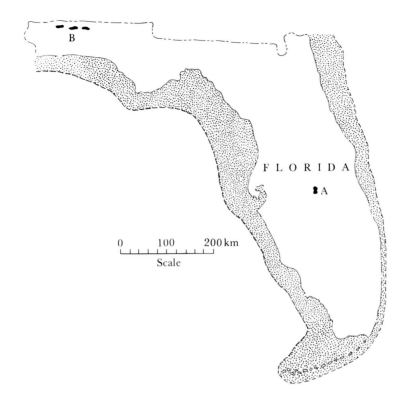

Figure 9.5. Major sea-level changes could devastate or expand Florida. If all the glacier ice melted, sea level would rise about 70 meters, reducing Florida to a group of three small islands in the northwest panhandle (B) and a single island at Iron Mountain in the center (A) indicated by black spots. A new glacial period attended by a possible sea-level drop of 120 meters would increase the state's land area by about 30 percent (stippled).

instrumental in creating, sustaining, and terminating an ice age, but the superintendent of the entire operation could be the sun.

THE PRESENT

If all the world's glaciers were suddenly melted, the results would be catastrophic. Florida would be reduced to a few tiny islands (Fig. 9.5), vast areas of heavily populated coastal lands would be submerged, and all the world's major harbors and many of its major cities would become gigantic aquaria, taken over by marine organisms accompanying the rise in sea level of 70 meters or so. The catastrophe and cost of such an unlikely event are almost beyond calculation or comprehension.

Current variations in glaciers, however, do cause small changes in sea level. The state of the antarctic ice budget has worldwide implications. If the budget is negative (wastage exceeding accumulation), sea level rises; if positive, sea level falls. Changes in the amount of antarctic ice are fully capable of causing a 1-meter change in sea level within a few centuries. Most studies of the Antarctic Ice Sheet suggest, however, that it has great inertia and is currently neither expanding nor shrinking perceptibly; the budget appears essentially to be balanced, with perhaps a slight tendency toward expansion.

Records clearly show that the smaller, nonpolar glaciers of the world react more quickly to environmental changes and have experienced significant episodes of advance and recession within the past century or two. The potential effects of this behavior upon eustatic shifts of sea level were recently evaluated by a workshop panel of scientists, with the conclusion that a rise of sea level by a good fraction of a meter was possible, even likely, within the next century owing to climatic amelioration, natural or artificial.

Imagine what a 1-meter rise of sea level would do to beaches, marinas, harbors, docks, seaside houses, lagoons, causeways, and countless other seashore features and structures. The rise of water level alone would be disastrous, and exposure to the energetic battering of waves and currents would wreak great havoc. The cost cannot be fully estimated, but it would be appalling. Longer-range ecological changes, mostly not immediately beneficial, would also occur. Whether this happens or not is out of our hands, and only time will tell.

To appreciate fully the virtues of glaciers, one has to live with them, and few of us have that opportunity. Even a casual observer can recognize, however, that they constitute an unusual element of the natural environment. Give them credit for sculpting some of the more remarkable areas of mountain scenery, and for further enhancing the scene if their vestiges linger in those areas. In the right places under the right conditions, glaciers can be strikingly beautiful. An ice fall can be dazzling and a water-filled moulin colorful. Internal reflections from melt figures within large, clear ice crystals rival a jeweler's display of diamonds. A sun-scalloped overhanging ice face is one of nature's finest sculptured products. There is much to see for those who have the eyes to see it.

Glaciers also render or have rendered useful services. They provide a most effective and economical means of storing water, especially for arid areas bordered by high mountains. Glaciers accumulate water in winter when it is least needed and release it in summer when it is most necessary, without anyone having to turn on the spigot. This they do without the building of a dam, clearing of a reservoir, or maintenance of any kind,

Figure 9.6. Large, cold ice sheets are good record keepers, or historians, particularly of happenings within the atmosphere. This shack was erected to shelter scientists collecting samples near Byrd Station in West Antarctica to trace the evolution of lead pollution in Earth's atmosphere caused by human industrial activity within historical times. The inscription on the shack walls says that the scientists built this shack themselves for their own benefit. (Photo courtesy of Hugh H. Kieffer.)

and without cost. Power companies in Switzerland, Scandinavia, and even to some extent in the United States depend on glaciers as a source of water, especially in winter.

Cold glaciers are good record keepers. The ice sheets of Greenland and Antarctica, where there is little or no melting, preserve the history of atmospheric variations and climatic change for the past 150,000 years in their layers of snow, firn, and ice. These glaciers know when the first atomic bomb was exploded and can tell the difference between early United States and Soviet nuclear tests. They know when lead tetraethyl was added to gasoline (Fig. 9.6) and when certain volcanoes exploded. They record variations in the flux of cosmic radiation impinging on the upper atmosphere through the amounts of beryllium isotopes (particularly ^{10}Be) preserved in the ice. They tell us about variations in the carbon dioxide content of the atmosphere over the past 100,000 years and that the amount of methane in the atmosphere has been increasing significantly in recent years. This is important because methane is being added more rapidly than carbon dioxide, and it contributes to the greenhouse warming effect.

Some easily detected chemical ions, especially chloride, sulfate, and nitrate, are preserved in glacier ice. Their variations reflect changes in environments, volcanic events, or human contributions to atmospheric pollution. Such data are obtained from cores from deep boreholes in Antarctica and Greenland. Two holes have been bored to the base of the Greenland Ice Sheet, the deepest at 2,037 meters. A deeper hole at Byrd Station in West Antarctica encountered bedrock at 2,164 meters. The Soviets are currently extending a borehole to an even greater depth at Vostok Station in East Antarctica. The age of the deepest ice is thought to be at least 125,000 years in Greenland and possibly as much as 200,000 years in Antarctica.

THE FUTURE

It is possible that the last Ice Age is not yet completed and that we may simply be living within an interglacial interval. Chances are reasonable that ice sheets may form once again and sweep over parts of the North American and Eurasian continents within the next 100,000 years or so.

The present world climate is about eight degrees Celsius cooler than what can be considered normal for the past 600 million years. A further drop of only three or four degrees might be enough to get new ice sheets started, and once they are under way a major climatic change may be required to stop them. The human race would not be swept from Earth's face by ice; however, related effects would be felt worldwide. The changes would be slow, and adjustments would be made over generations rather than in a single human lifetime.

It is interesting to speculate about some of the many possible consequences of a renewed glacial episode. Florida real estate, for instance, would really boom, and fortunately more of it would be available. Because growth of glaciers causes sea level to fall, perhaps by 100 meters or more, the dry-land area of the state would be increased by at least 30 percent (Fig. 9.5). An interesting issue would be whether the newly exposed land belonged to the federal government, the state, or the private landowners along the shore.

The world's major seaports would become landlocked, and new ports would have to be developed continually as sea level fell. If we were smart, facilities that could be moved would be constructed, so they could follow the receding shoreline. New Orleans would no longer have to pump its waste waters up into the ocean; they could run to the sea in natural channels. The Panama Canal would be high and dry. Far-reaching changes would occur in coastal settings around the world.

Because of the worldwide sea level drop and climatic changes, no place on Earth would escape untouched as the ice sheets developed. The effects of climatic change would be particularly great in arid and semiarid regions, where **pluvial** conditions would be established. The term *pluvial* is used in nonglacial areas for the cooler and moister condition that existed contemporaneously with episodes of Ice Age glaciation. A relatively small change in precipitation has a much greater effect in an arid region than in a well-watered area, so the imprints of pluvial conditions are most strongly expressed and keenly felt in arid or semiarid regions. It was under such conditions that many current desert areas once harbored large lakes and networks of streams. Droughts and starvation could become things of the past in central Africa under pluvial conditions.

Huge segments of the world's population would be displaced from northern regions into more habitable middle and tropical latitudes. This would generate contact and conflict with peoples already living in those areas. Some curious mixture of races could result. Even in more northerly areas not inundated by ice, living could become difficult. Recall the conditions earlier discussed characterizing areas peripheral to glaciers, the periglacial environment. Perennially frozen ground would create problems of water supply, sewage disposal, and foundations for roads and buildings. Siberians have long struggled with such matters, and U.S. Army Engineers suffered many surprises in dealing with them in Alaska during World War II. Their conclusion is: Learn to live with perennially frozen ground – don't fight it.

The impacts of renewed glaciation would not be all negative. Many parts of the globe now barely habitable could become delightful places in which to live. After the new glaciers receded, environmentalists would have an extensive area of pristine landscape to preserve. Many aspects of human depredation would have been erased, and plants and animals might develop interesting adaptations as they repopulated the newly deglaciated areas. Humankind would have suffered great material loss and physical stress, but such an experience might be just the thing to reinvigorate us as a biological species. Could it all happen? Yes, indeed!

Supplementary reading

ON GLACIERS AND THEIR BEHAVIOR

Paterson, W.S.B., *The Physics of Glaciers*, Second Edition, Pergamon Press, Oxford, 1981, 380 pp.
This is probably the best and most up-to-date single reference dealing with the basic mechanics of glacier formation, nourishment, structures, flow, and behavior available in the English language. Although it contains many pages and sections of rigorous physical analyses and derivations, a discerning nonprofessional reader can glean a world of useful, reliable information on glaciers from the intervening parts of the book.

Post, Austin, and LaChapelle, E.R., *Glacier Ice*, University of Washington Press, Seattle, 1971, 110 pp.
This large-size book provides an outstanding collection of superb air- and ground-based photos of glaciers and related features. The accompanying textual material is interesting, informative, and pertinent, like expanded figure captions.

Tyndall, John, *The Glaciers of the Alps and Mountaineering in 1861*, J.M. Dent, London (E.P. Dutton, New York), 1906, 274 pp.
This is an entertaining account of ramblings through the European Alps by one of Britain's outstanding physicists of the mid-nineteenth century as he applies his knowledge of basic physical principles to an early and enduring understanding of glaciers and glacier processes.

Schultz, Gwen, *Glaciers and the Ice Age*, Holt, Rinehart and Winston, New York, 1963, 128 pp.
This small book, written for laypersons, provides a good review of the basic aspects of glaciers and emphasizes their impact upon and relationships to human interests and activities.

Dyson, J.L., *The World of Ice*, Alfred A. Knopf, New York, 1962, 292 pp.
An older but highly readable book on glaciers and other natural bodies of ice, popular in its time.

Williams, R.S., *Glaciers: Clues to Future Climate*, U.S. Geological Survey, Reston, Virginia (Popular Publication) available through its eastern (Alexandria, Virginia) and western (Denver, Colorado) distribution centers, 1983, 21 pp.
This pamphlet provides a brief, popular, up-to-date summary on glaciers, causes of glaciation, glacial history, and climatic relationships.

Supplementary reading

Sugden, D.E., and John, B.S., *Glaciers and Landscape*, Edward Arnold, London, 376 pp.
This excellent, graceful book treats most of the items discussed in our book in an even more detailed and highly professional manner.

ON GLACIAL PROCESSES AND GLACIATION

Flint, R.F., *Glacial and Quaternary Geology*, John Wiley and Sons, New York, 1971, 892 pp.
A thoroughly scholarly, dependable, and detailed compilation and discussion of data, features, and processes related to glaciation and the worldwide history of glaciations. An excellent source book of facts and ideas.

Embleton, Clifford, and King, Cuchlaine A.M., *Glacial Geomorphology*, John Wiley and Sons, New York, 1975, 573 pp.
This book provides a highly detailed and thoroughly professional account of most features and matters treated in our book, and more. A good reference for everyone.

Eyles, N. (editor), *Glacial Geology: An Introduction for Engineers and Earth Scientists*, Pergamon Press, Oxford, 1983, 409 pp.
This is a modern, highly professional treatment of processes and products of glacial erosion and deposition. It is filled with useful and pertinent information, but one must dig for it.

Embleton, Clifford, *Glaciers and Glacial Erosion*, The Macmillan Press, London, 1972, 287 pp.
A small book providing a selection of parts or all of thirteen classical articles from the scientific glacial literature.

Ehlers, Jürgen (editor), *Glacial Deposits in Northwest Europe*, A. A. Balkema, Rotterdam, 1983, 470 pp.
Anyone interested in the details of glaciation in northwestern Europe will relish this volume, which treats mostly depositional features in Norway, Sweden, Denmark, West Germany, and the Netherlands. It is a pleasure to leaf through this handsome volume just to see the striking illustrations.

Ritter, Dale F., *Process Geomorphology*, Wm. C. Brown, Dubuque, Iowa, 1978, 603 pp.
Chapter ten of this book provides a nice survey and discussion of erosional and depositional features formed by glaciers. The writing is lucid and data are current.

Glossary

Terms in this glossary are purposely defined succinctly and solely with respect to their usage in this book. They may have somewhat different and more complex meanings in other contexts.

Ablation   The result of processes, primarily melting, that waste ice or snow from a glacier or snowfield.

Ablation Area   That part of a glacier's surface over which ablation (wastage) exceeds accumulation each year.

Ablation Debris   Mantle of unsorted rock debris that accumulates residually on the surface of melting, dirty glacier ice.

Ablation Till   Ablation debris that is ultimately deposited on the ground.

Abrasion   In erosion, the grinding down of rock surfaces.

Absorption   Transformation of radiant energy (light) as it passes through a substance.

Accordant Junction   Joining of two streams or valleys at a common level.

Accumulation   Deposition of ice, usually as snow, on the surface of a glacier.

Accumulation Area   Part of a glacier's surface over which more snow is deposited than is ablated each year.

Active Layer   Thin surficial layer, above perennially frozen ground, which undergoes freezing in winter and thawing in summer.

Albedo   That fraction of incident light reflected in all directions from a surface.

Alluvial Cone   Half-cone-shaped accumulation of rock debris below debouchment of a gully off a steep face; steeper, rougher, usually coarser, and smaller than an alluvial fan.

Alluvial Fan   Open-fan-shaped accumulation of rock debris at foot of a steep mountain slope, below debouchment of a canyon.

197

Alluvium   Unconsolidated gravel, sand, and finer rock debris transported and deposited by running water. (Adjective: alluvial).

Aphelion   Point at which an orbiting body is most distant from the body about which it is revolving.

Aplite   Fine-grained granitic rock, low in iron and magnesium.

Arête   Sharp, usually serrate, crestal rock ridge between two steep, glacially sculptured slopes.

Argon Dating   *See* Potassium-Argon Dating.

Armored Till Ball   *See* Till Ball.

Ash   *See* Volcanic Ash.

Atmosphere   Pressure of 1 million dynes per cm^2 at sea level.

Atom   Smallest unit of an element that retains its chemical identity; consists of nucleus and electron shells.

Backfill   Rock debris deposited into a backwater excavation earlier created by erosion, both commonly fluvial.

Banded Ogives   *See* Ogives.

Basal Gliding   Yielding of a crystal by interatomic slippage on a plane with basal crystallographic orientation.

Basal Slip   Slippage of a glacier over its bed.

Bedding   Layered structure of sedimentary rocks.

Belly Flowers   So small as to be observable only from a prone position.

Bernoulli's Theorem   Pertains to the decrease in pressure occurring with a homogeneous moving fluid as velocity increases.

Blowout   Hollow excavated by wind on the land surface, mostly in dunes.

Boulder Train   Fan-shaped deposit of glacial till, containing distinctive erratics derived from a localized upstream bedrock source.

B.P.   Abbreviation for "before the present."

Braided   Intricately branched and reunited pattern formed by channels of a wide, shallow stream flowing on a broad alluvial surface.

Calcite   Mineral composed of calcium carbonate ($CaCO_3$) with a rhombo-hedral lattice.

Calorie   Heat required to raise the temperature of 1 gm of water one degree Celsius.

Calving   Ice wastage by shedding of large ice blocks from a glacier's edge, usually into a body of water.

Carapace   *See* Ice Carapace.

Carbon-14   Radioactive isotope (^{14}C) of normal carbon (^{12}C) that disintegrates with a half-life of $5,660 \pm 30$ years. Used to measure geological ages between 300 and about 40,000 years B.P.

Celsius   Centigrade scale of temperature measurements.

Channeled Scablands   Peculiar topography created by huge ice-age floods, featuring abandoned, steep-walled channels with braided patterns, separated by flat-topped scablike remnants of dissected lava.

Cirque   Steep-walled, gentle-floored, semicircular topographic hollow created by glacial excavation high in mountainous areas.

Cirque Glacier   Small glacier lying wholly within a cirque.

Cirque Lake   Normally deep, clear-water lake lying in a largely bedrock basin on the floor of a cirque.

Clay   Rock or mineral particle smaller than 0.0039 mm; larger particles of clay-mineral family also qualify.

Col   Open, U-shaped pass across a high, narrow mountain ridge created by glacial erosion.

Cold Glacier   Below the pressure melting temperature of ice throughout. (Synonym: Polar Glacier.)

Columnar Jointing   Contraction cracks in volcanic rocks, which define parallel polygonal columns oriented perpendicular to a cooling surface.

Compound Cirque   Cirque with scalloped headwall resulting from coalescence of two or more formerly adjacent cirques.

Compound Valley Glacier   Composed of several ice streams.

Compressing Flow   Created by longitudinal compression in any reach of a glacier experiencing decreasing flow velocity; causes surfaceward movement of ice and an increase in thickness.

Concretion   Crudely spherical or irregularly rounded mass of firmly cemented rock material or segregated mineral matter within a larger body of sedimentary or fragmented volcanic rock.

Continental Ice Sheet   Sheet of ice of continental proportions, largely burying the underlying landscape and flowing outward in all directions.

Continental Shield   *See* Shield.

Convection   Transfer of heat within fluids by currents resulting from differences in density caused by differential temperature.

Cored Drumlin   Teardrop-shaped hillock of glacial till with a foreign core.

Coulee   Deep trenchlike gorge representing an abandoned overflow channel for glacier flood waters.

Couplet   Two adjacent sedimentary layers with different characteristics, rhythmically repeated.

Creep   Slow, essentially continuous, nonrecoverable deformation (movement) sustained by soils, rocks, and minerals (ice).

Crescentic Fractures   Small, hyperbolic, steeply inclined fractures created in brittle bedrock by friction from an overriding glacier.

Crevasse   Deep, steeply inclined, open fracture created in brittle surficial ice of a flowing glacier undergoing extension.

Crevasse Casts   Low, linear ridges of ground moraine squeezed into the bottom of a crevasse.

Crystal   Regular, solid, geometrical form bounded by plane surfaces, expressing an internal ordered arrangement of atoms. Aggregates of crystals may have irregular forms.

Crystal Fabric   Pattern of preferred orientation of crystals within a polycrystalline aggregate.

Crystal Lattice   Three-dimensional, regularly repeated arrangement of atoms within a crystal.

Crystalline Lithology   Mixed igneous and metamorphic rocks, or either separately.

Debris-Flow Deposit   Material laid down by debris flowage.

Debris Flowage   Relatively rapid downslope flow of a usually muddy mass of rocky debris under gravity; a type of mass movement. (Synonym: Mudflow.)

Declination   Difference, at any location, between the bearings to the magnetic pole and the geographic pole of Earth.

Density   Mass of a substance per unit volume.

Diamicton   Massive, poorly sorted, land-laid sedimentary accumulation containing angular rock fragments of all sizes in a fine (silt, clay) matrix.

Diatom   Microscopic, single-celled plant that grows in fresh or marine waters and secretes a complex silica skeleton.

Dike   Tabular body of intrusive igneous rock discordant to structure of host rock.

Discordant Junction   Joining of two streams or valleys at markedly different levels.

Discordant Tributaries   *See* Discordant Junction.

Dislocation   Defect within atomic structure of a crystal.

Dislocation Creep   Movement of dislocations through a crystal lattice under stress.

Distributary   Stream channel diverging from a trunk channel in a downstream direction.

Drift   All rock detritus carried and deposited in any mode by a glacier.

Drop Stones   Stones carried by and dropped from an iceberg or ice floe.

Drumlin   Low, streamlined ridge of glacial drift shaped by an overriding glacier. *See* Rock Drumlin, Cored Drumlin.

Dump Moraine   Ridge of glacial drift formed along a glacier's margin by rock debris dumped directly from the ice.

Dust Well   Small, vertical, cylindrical hole with thin bottom layer of dark silt formed by differential radiation melting on an ice surface.

Dyne   Force required to accelerate 1 gm of mass by 1 cm/sec per second.

Eccentricity   The distance between the two foci of an ellipse.

Electromagnetic Waves   Radiant energy waves propagated by electric and magnetic fields.

Electron   Atomic particle carrying a unit charge of negative electricity.

Emitted Radiation   Electromagnetic waves given off by any body.

End Moraine   Distinct accumulation of drift formed at the end of any glacier, including terminal and recessional positions.

Englacial   Within a glacier.

Equilibrium Line   Boundary between areas of gain and loss of ice on a glacier's surface during one year.

Equilibrium Zone   Irregular patchy band on a glacier's surface representing the equilibrium line.

Equinox   One of two times during a year when the sun is directly above Earth's equator.

Erosion   Removal of material, largely on Earth's surface, by any of a host of processes.

Erratic   Rock fragment, typically large, transported and deposited by a glacier.

Esker   Long, low, narrow, commonly sinuous ridge of glaciofluvial material deposited in a subglacial ice tunnel or an ice-walled channel.

Extending Flow   Created by extension within a glacier where flow is accelerating; causes thinning.

Extraglacial   Broad area peripheral to a glacier and dominated by processes, largely depositional, originating in the glacier.

Fabric   Pattern of preferred crystal orientation in polycrystalline glacial ice or of stones in glacial till.

Fault   Fracture surface or zone in Earth's crust along which block movements occur.

Firn   Loose, porous aggregate of small ice grains, more than one year old, transitional between snow and glacier ice.

Firn Line (Limit)   Lower edge of the firn mantle on a glacier at the end of an ablation season.

Fjord   Long, narrow, deep arm of the sea filling a glaciated coastal mountain valley.

Floe   Relatively large, usually flat, free-floating chunk of ice broken from the frozen surface of a river, lake, or sea.

Flood Plain   Strips of flat alluvial land bordering rivers; inundated by floods.

Flow Law   Governs the relationship between deformation of substances and the stress that produces it.

Flow Till   Till, usually of superglacial origin, that has been secondarily transported and emplaced by debris flowage.

Flute   Small, polished, elongate, scoop-shaped hollow on a rock surface eroded largely by fine debris transported by wind or water.

Fluted Moraine   *See* Ridged Moraine (preferred term).

Foci   The two points on the major axis of an ellipse, from which the form can be described by a string of fixed length greater than the separation of the points.

Foliation   Crude mineralogical or textural banding formed in rocks, primarily by solid-state metamorphism.

Fossil Ice Wedge   *See* Ice-Wedge Cast.

Freeze-Thaw Action   Sum of physical effects created by the change in volume between liquid and frozen water. (Synonym: Frost Action.)

Frequency   Of waves, oscillations per unit time.

Friction Cracks   *See* Crescentic Fractures.

Frozen Ground   Ground temporarily or perennially at a temperature below 0° C, wet or dry. (*See* Perennially Frozen Ground.)

Gendarme   Sharp rock pinnacle on an arête.

Geology   Science of the solid Earth.

Geothermal Flux   Amount of heat leaving Earth per cm^2 per second, average about 1.5 microcalories (0.0000015 calorie).

Geothermal Kame   Generally cone-shaped glaciofluvial deposit accumulated in a hole melted through a glacier by a subglacial hot spot, often a spring.

Giant Stairway   *See* Glacial Steps.

Glacial Drift   *See* Drift.

Glacial Geology   Study of the erosional and depositional work of glaciers.

Glacial Groove   Smooth, deep, usually straight furrow cut into bedrock by glacial abrasion; much larger than a striation.

Glacial Rebound   Elastic and isostatic upwarping of an area earlier depressed by a large body of glacier ice, now gone.

Glacial Steps   Succession of treads and high risers created by glacial erosion on the floor of a valley.

Glacial Till   *See* Till.

Glacier   Body of natural, land-borne ice that flows.

Glacier Budget   Amounts of accumulation and wastage of ice occurring in one year on a glacier.

Glacier Flea   Black springtail bug that lives within firn on glaciers.

Glacier Milk   Light-colored, glacially ground rock flour (silt) suspended in meltwater.

Glacier Table   Large, normally flat rock perched on a column of glacier ice because of differential melting.

Glacioeolian   Glacial material reworked by wind.

Glaciofluvial   Glacial material reworked by running water.

Glaciolacustrine   Glacial debris deposited in a lake.

Glaciology   Study of existing glaciers and, in a broader sense, of ice and snow.

Glide Plane   Crystallographic plane within a mineral, such as ice, along which displacement occurs during solid-state deformation.

Gradient   *See* Topographic Gradient.

Granite   Coarse-grained igneous intrusive rock composed largely of the minerals feldspar and quartz.

Gravity Wind   Sheet of cold air moving down a slope by gravity because of its greater density.

Groove   *See* Glacial Groove.

Groove Cast   Rounded, linear ridge of ground moraine molded into a groove engraved into the bottom of a glacier by a subglacial bedrock knob or fixed boulder.

Ground Ice   Subsurface body of pure ice, usually within frozen ground.

Ground-Ice Wedge   Wedge-shaped, near-vertical body of laminated ground ice, pointed downward; grows by annual increments.

Ground Moraine   More or less continuous mantle of till with gently undulating surface; deposited principally under an ice sheet or along the edge of a steadily receding glacier.

Gumbotil   Mostly dark-colored, clay-rich, sticky (wet), thick soil layer formed by long weathering of till under moist conditions, poor drainage, and low topographic relief.

Halite   Common salt, a mineral (NaCl).

Hanging Valley   Tributary valley with a floor distinctly higher at junction than the trunk valley.

Hardness   Resistance of a mineral to scratching.

Hardness Scale   Nonlinear scale based on hardness of ten well-known minerals; from talc at 1, through apatite at 5, to diamond at 10.

Headwall   Steep cliff backing a cirque.

Hoarfrost   Deposit of delicate skeletal ice crystals primarily by sublimation from water vapor upon a surface cooled by radiation loss.

Holocene Epoch   Past 10,000 years; succeeds the Pleistocene.

Horn   High, sharp, steep-sided pyramidal peak, sculptured by cirques working headward from several sides.

Hydrodynamics   That part of hydromechanics dealing with forces producing flow.

Hydrogen Bonding   Chemical bond in which a positive hydrogen atom forms a bridge between two electronegative oxygen atoms in two different ice or water molecules.

Ice Age   Period when large sheets of ice inundate nonpolar continental areas.

Iceberg Large, massive chunk of glacier ice floating in a water body or stranded on its shore.

Ice Cap Sheet of glacier ice, of less than 50,000 km², capping an upland and spreading out radially.

Ice Carapace Sheet of ice covering upper part of a high, isolated peak.

Ice-Contact Feature Accumulation of glacial drift with a configuration largely determined through deposition in contact with glacier ice.

Icefall Like a waterfall but composed of glacier ice.

Ice Island Large, detached chunk of shelf ice floating in midst of sea ice.

Ice Lobe Tongue-shaped mass of glacier ice projecting from a larger body, usually a sheet, cap, or carapace.

Ice-Rafted Rock debris carried by an iceberg, floe, or other floating mass of ice.

Ice Sheet Very large, thick sheet of glacier ice, covering areas greater than 50,000 km², largely submerging the subglacial topography, and flowing generally outward in all directions.

Ice Shelf Large sheet of glacier ice shoved out over the sea floor from a land-based source; likely to be partly afloat.

Ice Stream Stream of ice flowing down a valley or a current of high velocity within an ice sheet. (Synonym: Valley Glacier.)

Ice Wedge *See* Ground-Ice Wedge.

Ice-Wedge Cast Filling of loose rock detritus, commonly sandy, in the space vacated by a melted ice wedge.

Ice Worm Black segmented worm, 2 to 3 cm long, that lives in glacier ice and firn.

Igneous Rock One composed principally of minerals that crystallized from a melt (magma); may be glassy, if chilled, or fragmental, if explosive in origin.

Illinoian Glaciation Next-to-last phase of Pleistocene glaciation in North America; named from extensive drift deposits in that state.

Impact Creep Movement of loose fragments along the ground under impact of saltating windblown sand grains.

Indicator Fan Fan-shaped till mass containing fragments of a highly distinctive rock plucked from a bedrock source by a glacier.

Indicator Stone Erratic of a highly distinctive rock derived from an identified bedrock exposure. *See* Erratic.

Inset Ice Stream Tributary stream of ice inset into the surface of a compound valley glacier.

Insolation   Radiated energy received from the sun.

Interglacial Forest   Grew in a deglaciated area and was then overrun by re-advancing ice.

Interglacial Interval   Warmer and possibly drier period separating two major glacial phases.

Interlobate Moraine   Marginal morainal complex formed between two closely adjacent ice lobes.

Isotope   Species of a chemical element that has one or more neutrons in its atomic nucleus above what is normal.

Joint   Plane fracture surface cutting through a rock mass without displacement.

Jointing   System of parallel joint sets within a rock body.

Jökulhlaup   Outburst flood of glacial origin; common in Iceland.

Juxtaposed Ice Streams   Two or more side-by-side ice streams in a compound valley glacier that extend from the surface to the floor.

Kame   Steep-sided hillock or ridge of glaciofluvial debris deposited in contact with glacier ice.

Kame Terrace   Tread of glaciofluvial material deposited along the margin of a glacier, lying above a riser formed by deposition against the former ice edge.

Kansan Glaciation   Next-to-earliest of the four Pleistocene glaciations in North America.

Kettle   *See* Kettle Hole.

Kettle Hole   Normally bowl-shaped, topographically closed depression within glacial drift formed by melting of a large chunk of partly or completely buried glacier ice, usually filled with water.

Kettle Valley   Kettle hole with the configuration and dimensions of a small valley.

Kinematic Wave   Wavelike perturbation of a glacier's steady-state flow that moves outward at a velocity several times normal.

Knob-and-Kettle Topography   Undulating surface on glacial deposits characterized by knobs separated by bowl-like depressions.

Laminar Flow   Fluid movement in which the flow lines remain separate and unchanged with time at every point.

Lateral Moraine   Accumulation of glacial drift along the lateral margins of a valley glacier, remaining as a ridge or embankment upon glacier recession. Also the surficial accumulation of ablation debris on the margin of an existing valley glacier.

Lattice *See* Crystal Lattice.

Laurentide Ice Sheet Continental ice sheet covering central and eastern Canada during Pleistocene Ice Age.

Lithification Changes created by desiccation, compaction, crystallization, and cementation that convert unconsolidated debris into a rock.

Lodgment Till Unsorted rock debris deposited by a plastering-on process along the bed of an active glacier; normally compact, rich in fine particles, and with oriented stones.

Loess Deposit of wind-blown dust; homogeneous, massive, and fine-grained.

Loess Children Small concretionary masses within loess bearing a resemblance to all or part of a human child.

Longitudinal Septum Longitudinal band of strongly and complexly foliated, somewhat dirty ice within a valley glacier extending downstream from a rock rib or bastion in the face of an icefall.

Marble Monomineralic rock composed of calcite, formed by metamorphism of limestone.

Marginal Moraine Moraine formed along the margin of an ice sheet, cap, or lobe, not appropriately described as lateral or terminal.

Mass Inertia of a body being accelerated by gravity; equals weight divided by the acceleration of gravity.

Mass Creep Slow movement downslope by a body of loosened rock debris. *See* Creep.

Mass Movement Downslope flow or sliding of a mass of rock or rock debris, usually as a unit.

Mass Spectrometer Laboratory instrument that separates different chemical elements and their isotopes on the basis of unit specific mass.

Medial Moraine Longitudinal stripe of ablation debris on a glacier's surface marking the trace of a septum of dirty ice formed at the junction of two adjacent ice streams in a compound glacier.

Megaflute *See* Flute.

Melt Figure Disk-shaped opening, a few millimeters in diameter, containing water and vapor, formed by melting under solar radiation within a clear ice crystal, oriented parallel to the basal crystallographic plane. (Synonym: Tyndall Figure.)

Melt Till Rock debris laid down beneath a glacier through melting of the basal, debris-rich ice layer.

Metamorphic Rock   Derived from preexisting rocks primarily by solid-state changes in mineralogy and structure through heat, pressure, stress, and changes in chemical environment, without extensive melting.

Mineral   A natural inorganic solid of fixed chemical composition and definite crystal structure.

Molecular Scattering   Emission of radiant energy owing to electron oscillation. (Synonym: Rayleigh Scattering.)

Molecule   Smallest particle of a substance retaining all chemical and physical characteristics of the substance.

Monomineralic Rock   Composed of a single mineral.

Moraine   Distinct ridge or assemblage of ridges and mounds representing a concentrated accumulation of drift of any kind, laid down directly from a glacier at a position of enduring stability.

Moulin   Roughly cylindrical, near-vertical hole up to 30 meters deep in a glacier's surface.

Mudflow   Rapidly flowing mass of mud commonly containing coarse rock fragments; a type of mass movement. Term also applies to resulting deposits. *See* Debris Flowage.

Nanometer   One billionth of a meter.

Nebraskan Glaciation   First phase of Pleistocene glaciation in North America.

Neutron   Electrically neutral atomic nuclear particle with a mass nearly the same as that of a proton.

Névé   French term for old granular snow; called *Firn* in German.

Newtonian Fluid   Yields to shear stress at a constant ratio.

Nunatak   Eskimo term for an island of bedrock projecting through a surrounding mass of glacier ice.

Ogives   Rhythmically repeated, light and dark banding within a glacier or a swell-and-swale configuration on its surface, convexly curved downstream, that forms below some icefalls.

Outlet Glacier   Ice stream or lobe normally flowing rapidly outward from a larger body of ice, usually a sheet or cap.

Outwash   Glaciofluvial material deposited outward from a glacier by meltwater streams.

Outwash Plain   Broad, gently sloping plain of outwashed glaciofluvial debris.

Outwash Terrace   Step-and-riser form created in outwash deposits.

Overburden Pressure   Confining pressure exerted by the overlying weight of ice.

Overlapping Spurs   Ridges between tributaries to a steep-walled trunk canyon that appear to overlap from opposite sides as viewed in profile.

Oxygen-Isotope Ratio   Most commonly, the ratio of the ^{18}O isotope of oxygen to oxygen-16 in any fluid, vapor, or solid state of water.

Paleotrimline   Sharp line between trees of distinctly different age created by old glacier advance into forested terrain.

Parameter   Any measurable characteristic of a system that can be expressed numerically and is constant or a function of a related system of variables.

Paternoster Lakes   String of glacially excavated, possibly partly dammed lakes on a valley floor connected by a trunk stream, resembling beads on a rosary chain.

Patterned Ground   *See* Stone Polygon, Stone Stripes.

Perched Water Table   Ground water, not part of a confined aquifer, held above the normal ground-water table by an impervious zone.

Perennially Frozen Ground   Continually below 0° C, year after year. (Equivalent: Permafrost.)

Perfectly Plastic Substance   Shows no deformation under increasing shear stress up to a point, after which it deforms continuously without rupture or further increase in stress.

Periglacial   Cold, rigorous environment characterizing the area peripheral to a large Pleistocene glacier, and the processes and products found there.

Perihelion   Point at which an orbiting body is closest to the body about which it is revolving.

Peripheral   Extraglacial zone immediately along the ice edge.

Permafrost   *See* Perennially Frozen Ground.

Permeability   Capacity of a porous material to transmit a fluid.

Photon   Unit of radiant energy moving with the velocity of light.

Piedmont Glacier   Sheet of ice formed by one or more ice streams spreading out on a flatland at base of mountains.

Piedmont Lake   Water body in basin vacated by a piedmont glacier.

Pitted Outwash   Outwash surface indented by numerous kettle holes.

Plate Tectonics   Movement and interaction of huge crustal plates covering Earth's entire surface and afloat in the viscous layer of Earth's mantle.

Pleistocene Epoch   Approximately the past 1.6 million years of Earth's history; essentially equivalent to the last great Ice Age.

Plucking   Glacial erosive process whereby sizable blocks of rock are loosened, picked up, and carried away by glaciers. (Synonym: Quarrying.)

Pluvial   Conditions of greater precipitation and generally cooler temperatures existing in nonglacial areas during a glacial period.

Poise   Unit for measuring dynamic viscosity, involving force, time, and area.

Polar Glacier   *See* Cold Glacier.

Polycrystalline   Composed of many individual crystals.

Polygon   *See* Stone Polygon.

Potassium-Argon Dating   Radiometric method of determining age of rocks or minerals containing radioactive potassium-40, which decays to argon-40 at a known rate.

Pressure Melting   Melting of ice where pressure is great enough to lower its melting point below the ambient temperature.

Pressure Melting Temperature   Temperature at which ice will melt under a specified pressure.

Proglacial   Area lying adjacent to and usually in front of a glacier.

Proton   Elementary particle of an atomic nucleus carrying unit positive electric charge.

Pseudo-Igneous   E.g., ice, which forms like an igneous rock by crystallization from a molten state.

Pseudo-Viscous   Solid substances that behave in part or temporarily as though they were viscous fluids.

Push Moraine   Ridge of rock debris shoved up along the edge of an advancing glacier.

Quarrying   Process of glacial erosion involving plucking.

Quasi-Plastic   Solid-state deformation resembling, but not duplicating, a plastic behavior.

Radiation   Transmission of energy in the form of electromagnetic waves.

Radiocarbon   *See* Carbon-14.

Rattails   Small, low protuberances on a glacially scoured rock surface, with blunt fore end and a tapering tail several centimeters long.

Rebound   *See* Glacial Rebound.

Recessional Moraine End moraine built during a pause or slight readvance of a receding glacier.

Refraction Change in direction of a light ray passing from one medium to another of different density.

Regelation Twofold physical process of ice melting under applied pressure and refreezing upon release.

Regelation Layer Thin layer of refrozen, granular, laminated, usually dirty ice at the base of a glacier.

Ridged Moraine Moraine having distinctive surface patterns of linear ridges and flutes. (Also called Fluted Moraine.)

Rime Deposit of dense ice formed by freezing of droplets of supercooled fog upon a solid surface.

Riser Steep element of a step rising to or from a more horizontal tread.

Roche Moutonnée Asymmetrical, glacially eroded bedrock hillock with smooth, abraded, more gentle flank on the upstream (stoss) side and steep, irregular, ragged flank on the lee side.

Rock Naturally formed, consolidated aggregate of mineral crystals or rock fragments.

Rock Debris Surficial broken-up or decomposed rock.

Rock Drumlin Streamlined, teardrop-shaped bedrock hillock sculptured by glacial erosion.

Rock Flour Fresh, clay- and silt-sized rock and mineral particles ground by glacial abrasion.

Saltation Transport mode of wind or water in which particles move by a succession of hops.

Sand Rock or mineral particles with a diameter between 0.0625 and 2.0 mm.

Sandblasting Erosive action of a stream of windblown sand particles driven against an object.

Sapping Undercutting of a steep face by ground-water seepage, glacier quarrying, or accentuated weathering at its base.

Sawtoothed Ridge A serrated rock divide.

Scab *See* Channeled Scabland.

Scour and Fill The deepening of a stream channel by scour during waxing phase of a flood and subsequent refilling as the flood wanes.

Sea Ice Ice floating on the surface of a sea, formed by freezing of ocean water.

Sediment Unconsolidated rock detritus that is transported and deposited by one of several different media.

Sedimentary Rock Consolidated sediment, normally stratified (layered).

Serac Large, irregular, sharp-crested and steep-sided block of glacier ice shaped by calving between crevasses in an icefall.

Serrate Ridge *See* Sawtoothed Ridge.

Shear Stress Component of stress tangential to a plane.

Sheeting Jointing parallel to a rock surface resulting from expansion or release of pressure.

Shield Large area of old crystalline rock, usually largely Precambrian in age, forming the core of a continent.

Sickle Troughs Crudely crescentic depressions on glaciated rock surfaces.

Sill Shallow, submerged ridge across which water must flow to enter or escape from a deeper basin.

Silt Rock or mineral particles with a diameter between 0.0625 and 0.0039 mm.

Snow Aggregate of delicate skeletal ice crystals formed in the atmosphere by sublimation of water vapor.

Snowline Lower boundary of a continuous snow cover on a landscape.

Solar Constant Rate at which radiant solar energy is received at the top of Earth's atmosphere; mean value 1.94 calories per minute per cm^2.

Solid Flow Permanent yielding by flow of a solid mass owing to rearrangement between its particles or atoms.

Sorting Arrangement of particles by size.

Spalling Breaking off of flakes from a rock surface.

Specific Heat Ratio of the quantity of heat required to raise the temperature of a body by one degree to that required to do the same to an equal mass of water.

Spectrum Ordered array of intensity values of some physical parameter, such as wavelengths of light.

Spur Subordinate ridge extending from a primary ridge.

Stagnant Glacier A glacier, or part of one, that has stopped moving and is wasting away in place.

Stone Polygon Polygonal border of stones around a central area of much finer debris, formed by freeze–thaw action.

Stone Stripes Narrow, linear bands of coarse, sorted stones separated by wider bands of finer debris extending down slopes; created by freeze–thaw action and creep.

Stress Force per unit area acting on any surface within a solid.

Striae *See* Striations.

Striations Linear, finely cut parallel scratches inscribed on a rock surface by debris carried in basal ice of a moving glacier.

Subglacial The realm beneath a glacier.

Superglacial On the top surface of a glacier.

Superimposed Ice Stream Tributary ice stream riding on the surface of its trunk glacier.

Surge Relatively short-lived episode of greatly accelerated flow within a glacier.

Suspension Transport mode experienced by fine particulate debris supported by upward-directed turbulent currents in a fluid.

Swell-and-Swale Ogives *See* Ogives.

Tectonic Activity Deformation of Earth's crust by forces originating within the planet.

Temperate Glacier At the pressure melting temperature of ice throughout. (Synonym: Warm Glacier.)

Terminal Moraine Outermost end moraine deposited during a major episode of glacier advance.

Terrace Landscape feature resembling a step and consisting of a long, relatively narrow tread above a steep riser.

Terrain Large area of similar landscape.

Till Ill-sorted, mixed fine and coarse rock debris deposited directly from glacier ice.

Till Ball Coherent chunk of till, rounded by fluvial transport, incorporated within glaciofluvial deposits. Outer surface may be armored by adhered stones.

Till Sheet Essentially continuous mantle or layer of till. (Synonym: Ground Moraine.)

Tillite Firmly cemented till, usually aged.

Topographic Gradient General slope of land surface, expressed in degrees or height change over distance (meters per kilometer).

Traction   Transport mode in which particles of rock debris are rolled, pushed, or slid across the ground by wind or water.

Transected Divide   Topographic divide that has been breached by some erosive process, commonly a transection glacier.

Transection Glacier   One that flows across a topographic divide because of its great thickness.

Tread   The smooth, nearly flat element of a terrace.

Trimline   Sharp line between vegetated (forested) and essentially denuded terrain marking outer limit of a recent glacier advance.

Tunnel Valley   Small valley or trough cut into drift or bedrock by a subglacial stream flowing in an ice tunnel.

Turbidity Current   Muddy current within a body of clearer fluid that flows independently because of gravity, owing to greater density created by fine, suspended particulate matter.

Turbulent Flow   Fluid movement in which the flow lines go in all directions with great variations in velocity and orientation.

Tyndall Figure   *See* Melt Figure.

U-Shaped Valley   A wide, open valley with steep walls and nearly flat floor, carved by a valley ice stream.

Valley Train   Long, narrow accumulation of outwash material extending downstream from the stabilized front of a valley glacier.

Vapor Transfer   Movement of solid substance (ice) in the vapor state from a site of evaporation to a site of deposition.

Variable   Any measurable statistical quantity that changes independently or dependently with other quantities.

Varve   Glaciolacustrine sedimentary couplet of contrasting summer- and winter-laid material, representing one year of deposition.

Vector   Physical quality that indicates both direction and magnitude; can be represented by a barbed arrow of scaled length.

Vein   Secondary filling of mineral matter along a fracture in a rock.

Ventifact   Stone with a surface modified (polished, pitted, fluted, grooved, or faceted) by wind blasting.

Viscoplastic   Plastic substance that behaves somewhat like a viscous fluid.

Viscosity   Internal resistance to flow offered by a substance.

Volcanic Ash   Unconsolidated, fine, explosively fragmented volcanic material of particle diameter less than 4 mm.

V-Shaped Canyon   Stream-cut canyon with narrow floor and steep walls forming a V-shaped cross profile.

Warm Glacier   *See* Temperate Glacier.

Washboard Moraine   Moraine with parallel-running swales and swells resembling subdued waves.

Wastage   All processes, predominantly calving and melting, by which snow and ice are removed from a glacier.

Wastage Area   *See* Ablation Area.

Water Table   Subsurface level below which all openings in rocks are filled with water.

Wavelength   Distance between comparable points on successive waves in a wave train.

Weathering   The chemical alterations and physical disintegration produced in surficial materials of Earth's crust through interaction with the atmosphere, biosphere, and hydrosphere.

Whaleback   Smooth, glacially sculptured bedrock knob of modest size resembling the back of a sounding whale.

Wisconsinan Glaciation   Fourth and last glacial phase of the Pleistocene Ice Age in North America; ended about 10,000 years ago.

Index

Index

Index

Index

Index